Balkanization and Global Politics

Ever-increasingly, countries, states and regions are voicing a desire to be autonomous through a process of balkanization. This book explores the historical emergence, interdisciplinary application and current sociospatial reasons why more places are seeking self-governance around the world. The spatialization of balkanization is particularly addressed in terms of destruction and renewal through a detailed sociopolitical interrogation of architecture and the urban, including their changing symbolic and functional forms.

The book offers a reworking of the concept of balkanization through a reflective and critical analysis. The particular attention on the city of Belgrade, including the 1990s dissolution of Yugoslavia through specific case study focus of Croatia, Bosnia-Herzegovina and Serbia, provides insightful connections between balkanization, violent remaking and global politics. Against the detailed historical overview and prevailingly negative understanding of balkanization, a more positive instatement of balkanization for purposes of inclusivity and coexistence is also presented.

The book will be relevant to academics and students interested in spatial politics. The broad analysis will appeal across disciplines such as Geography, Politics, Architecture and Urban Studies.

Nikolina Bobic is a Lecturer in Architecture at the School of Art, Design and Architecture, University of Plymouth, UK. Her research delves into the historical and theoretical concepts to do with spatial violence, control and balkanization.

Routledge Studies in Urbanism and the City

This series offers a forum for original and innovative research that engages with key debates and concepts in the field. Titles within the series range from empirical investigations to theoretical engagements, offering international perspectives and multidisciplinary dialogues across the social sciences and humanities, from urban studies, planning, geography, geohumanities, sociology, politics, the arts, cultural studies, philosophy and literature.

Indigenous Rights to the City
Ethnicity and Urban Planning in Bolivia and Ecuador
Philipp Horn

City Branding and Promotion
The Strategic Approach
Waldemar Cudny

Citizenship and Infrastructure
Practices and Identities of Citizens and the State
Edited by Charlotte Lemanski

Tourism and Everyday Life in the Contemporary City
Edited by Thomas Frisch, Christoph Sommer, Luise Stoltenberg and Natalie Stors

Balkanization and Global Politics
Remaking Cities and Architecture
Nikolina Bobic

Peri-Urban China
Land Use, Growth and Integrated Urban Rural Development
Li Tian

For more information about this series, please visit: www.routledge.com/series/RSUC

Balkanization and Global Politics

Remaking Cities and Architecture

Nikolina Bobic

Routledge
Taylor & Francis Group
LONDON AND NEW YORK

First published 2019
by Routledge
2 Park Square, Milton Park, Abingdon, Oxon OX14 4RN

and by Routledge
605 Third Avenue, New York, NY 10017

First issued in paperback 2020

Routledge is an imprint of the Taylor & Francis Group, an informa business

British Library Cataloguing-in-Publication Data
A catalogue record for this book is available from the British Library

Library of Congress Cataloging-in-Publication Data
Names: Bobic, Nikolina, author.
Title: Balkanization and global politics / Nikolina Bobic.
Description: First edition. | Abingdon, Oxon ; New York, NY : Routledge, 2019. | Series: Routledge studies in urbanism and the city | Includes bibliographical references and index.
Identifiers: LCCN 2018054356 | ISBN 9781138061835 (hbk : alk. paper) | ISBN 9781315162065 (ebk)
Subjects: LCSH: Balkan Peninsula—History—Autonomy and independence movements. | Architecture—Political aspects—Former Yugoslav republics. | Former Yugoslav republics—Politics and government.
Classification: LCC DR1316 .B634 2019 | DDC 949.703—dc23
LC record available at https://lccn.loc.gov/2018054356

ISBN 13: 978-0-367-73081-9 (pbk)
ISBN 13: 978-1-138-06183-5 (hbk)

Typeset in Times New Roman
by Apex CoVantage, LLC

To Mira and Nikola

Contents

Abbreviations

CIAM – International Congresses of Modern Architecture
DU – Depleted Uranium
EBRD – European Bank for Reconstruction and Development
EU – European Union
FB-H – Federation of Bosnia and Herzegovina
FRY – Federal Republic of Yugoslavia
GPS – Global Positioning System
HDZ – Croatian Democratic Union
HV – Croatian Army
HVO – Croatian Military Defence
ICTY – International Criminal Tribunal for Yugoslavia
IMF – International Monetary Fund
IPAP – Individual Partnership Action Plan
JNA – Yugoslav People's Army
KLA – Kosovo Liberation Army
MIME – Military-Industrial-Media-Entertainment
NATO – North Atlantic Treaty Organization
NDH – Independent State of Croatia
ORA – Youth Labour Brigades
RSK – Republic of Serbian Krajina
RTS – Radio Television of Serbia
SFRY – Socialist Federal Republic of Yugoslavia
UK – United Kingdom
US – United States
UN – United Nations
UNESCO – United Nations Educational, Scientific and Cultural Organization
UNHCR – United Nations High Commissioner for Refugees
UNMIK – United Nations Interim Administration Mission in Kosovo
USSR – Union of Soviet Socialist Republics
VCM – Vinyl Chloride Monomer
WWI – World War I
WWII – World War II

Figures

Note: All photos are author's own.

Preface

The historically received understanding of balkanization acquired its negative connotations from the experience of fragmentation and division as imposed through violence in response to sociopolitical instability and conflicts in the Balkans. This book explores these processes of violent remaking (destruction and renewal) through a detailed interrogation of architecture and the urban in terms of their changing symbolic, ideological and functional forms. Broadly, this is done through a reflective and critical engagement with the history of the Balkans understood through the discourse of Balkanism and balkanization. Yet there is another, currently neglected, tradition of balkanization from the very same territory that gave rise to the term as a geopolitical synonym for political fragmentation, division and hostility. This lesser known tradition provides an inclusive and hopeful vision, which this book investigates in detail as it has manifested in the recent history of Belgrade as well as in the more distant past of the former Yugoslav context.

The focused analysis helps in understanding broader emergent patterns of sociospatial polarization across various scales, and in respect to global geoeconomic and geopolitical restructuring. This is particularly important because balkanization is becoming a significant urban and geopolitical pursuit in contemporary times; countries, cities and regions are ever-increasingly voicing the desire for independence. Drawing connections between balkanization, economics, law, military, media and technology is to gain awareness of, and to engage with, the emerging implications of spatial remaking and global politics.

Acknowledgements

The embryo for this book was generated during my dissertation at the University of Sydney, in the currents of a long-standing interest in the political implications of architecture and the urban. Some of the thinking was tested in the following publications:

"Belgrade (de)Void of Identity: Politics of Time, Politics of Control, Politics of Difference," *Cultural Dynamics* 29(1–2): 2017, 3–22.
"War Machine: Media and Technology during Operation Allied Force," *Fabrications* 25(3): 2015, 398–416.
"New Belgrade After 1999: Spatial Violence as De-socialisation, De-Romanisation, and De-historisation," *Architectural Theory Review* 19(3): 2014, 355–375.
"Belgrade in Formation(s): Dobrovic's Generalstab Complex," *Fabrications* 21(1): 2012, 6–27.

However, my deep immersion in the discourse of Balkanism and balkanization at the intersection of architecture and the urban occurred during the research and writing of the present manuscript. It has been a complex yet affirmative process; it has also been my period of introspection and growth.

I have become indebted to many people and institutions through the course of this academic undertaking. The initial support for and belief in my research came from Dr. Katharine Willis; for her guidance I am grateful. I also thank Andy Humphreys for his understanding and affordance of breathing space between my teaching workload and the writing of this manuscript. I would also like to recognize the University of Plymouth and the help it has extended in the form of funding for field research trips and teaching relief.

Special thanks go to staff at the many libraries, museums and archive centres in the former Yugoslav context for facilitating primary research access. George Davis drew all the maps featured in the book, a contribution that I very much appreciate. Thanks also go to Routledge for seeing value in this topic, to the unknown reviewers of the manuscript proposal and particularly to the wonderful commissioning editor Faye Leerink and the ever patient and understanding editorial assistant Ruth Anderson. My sincere thanks also to David Kelly for his thoughtful comments and editing services.

To all my extraordinary teachers past and present – too many to name – I owe fondest gratitude and thanks. The incredibly constructive feedback provided by Dr. Martin Coward and Dr. Chris L. Smith helped reshape and tighten the manuscript, as well as prompted thinking for my future research – thank you.

Most of all, thank you to my immediate family – you know who you are – for being you and, for your love and light which always guide me and help me evolve in ways I could never imagine. I love you beyond words!

1 Setting the framework

Balkan(ization) and global politics

Framing Balkanism and balkanization

Balkanism

In a world where countries, states, cities and regions are increasingly voicing the desire for independence from the nation or union they are a part of, it seems that balkanization – that is, geopolitical fragmentation – is manifesting on an ever-wider scale. At the time of writing, the push towards balkanization was most recently seen in the vote by the UK to part ways with the European Union (EU). On a country and regional scale, some other examples include Scotland's close-run attempt to secede from the UK, and in Spain the impetus to separate in the Basque Country and Catalonia. In the current political and economic climate, balkanization seems to a global concern. However, to understand balkanization, there is a need to first frame the discourse of Balkanism.

The Balkans is perceived through the lens of violent imaginary of racial and ethnic hatreds; *Balkanism* being a *mode of identification* prompted by the expulsion of the Ottoman Empire from the Balkans in the 19th Century and the geopolitical balkanization through fragmentation and reorganization of the region that flowed from that. More broadly, Balkanism is also a signifier of violence associated with the Balkan War in 1912–13 and the beginning of WWI. The construct being that Balkanism is opposed to supposed Western civilization. Coined by Maria Todorova in 1997, the term Balkanism emerged due to this geographical zone not fitting within the European Occidental ideal, yet not being Oriental enough to be called Oriental.[1] In other words, Balkanism poses an identification problem; its peripheral location and its social, cultural and spatial behaviour are neither Occidental nor Oriental enough, resulting in it being seen as some form of agitator, not fitting a binary 'us' and 'them' construct.

The study of Balkanism not only emerged in the Balkans but is also – according to Dušan Bjelić – an "intellectual export industry of the Balkans."[2] Balkanism encompasses that which has been associated historically with the region as well as critical responses to that imaginary.[3] Unlike Edward Said's Orientalism, which addressed the need to refrain from certain perceptions and identities about the Orient,[4] the literature on Balkanism, in contrast, affirms constitutive differences and paradoxes for the sake of the Balkans' representational concreteness in terms of the relationship between the Balkans and the West.[5] Also, the presence of colonization is less obvious and direct, unlike that in the Orient.

Balkanism unsettles what is perceived as solid. The Balkans have been described as a zone of transition and passing through,[6] and "a bridge between stages of growth," thus invoking labels such as "semideveloped, semicolonial, semicivilized, semioriental."[7] For such a liminal and complex series of identifications, Balkanism outside the borders of the eponymous region tends nevertheless to be seen within the fixed and simplistic framework of a zone of irreversible violence and dissent spurred by fragmentation. That reductionism is the common currency when describing peripheral and Other zones is, unfortunately, an unremarkable feature of the contemporary climate.

The need to legitimize homogeneity through adoption of Western values was seen in the renaming of the Balkans as the Western Balkans in 1998.[8] The renaming was contingent on balkanizing this geopolitical region, from the perspective that only those states still associated with the Balkanist rhetoric need to be grouped together and called the Western Balkans. The identification is, thus, not used for geographical purposes or geopolitical ones but rather as a new way of labelling all those areas that resist a firm identity. Though, to deploy Stephen Graham's thinking, geopolitics is a flat discourse, as it largely focuses on the plane across, and predominantly disregards the vertical plane that cuts through.[9] Such a focus on horizontality alone facilitates dismissal of the politics in the change of name, politics geared towards an insistence that this rogue zone attain Western stability and progress.

To think Balkanism in terms of the politics of the Western Balkans is, for Rastko Močnik,[10] encompassing of both horizontal and vertical dimensions. Horizontality is inclusive of antagonisms found between various ethnicities and states within the Balkans, while the vertical plane is associated with various Western contractual impositions to do with required collaboration between all the Yugoslav belligerents and the EU.[11] However, to address the region only in terms of horizontal and vertical dimensions is reductive, particularly when these dimensions come into contact with questions of violence. To plot the politics against cartesian coordinates is to disregard the national and global forces acting upon the region as well as the way in which changes are contingent on the time at which they occur.

Balkanization

The imposition of balkanization for purposes of attaining stability emerged when the Balkans and its borders were reorganized during the 1878 Congress of Berlin. Very broadly, both of the terms – Balkanism and balkanization – are largely synonymous with dissolution, division and lack of dialogue and cohesiveness. However, to affix this definition to only the Balkans is to divide by distinguishing areas and peoples who violate from those who defend. The association of the Balkans with violence seems to have a distinctly reductive historical framing in that such connotation 'logically' leads to balkanization linked to Balkanism, and thus associated with a violent parcelization of a territory into smaller zones claiming ethnic homogeneity.

A reluctance to relinquish the villainous image of the Balkans thus indirectly determines the knowledge we have, indeed the very construct, of this geographical region. Despite the commonalities – balkanization being associated with fragmentation and at times Balkanist prejudices – they cannot be reduced to one another since balkanization does not always coexist and overlap with Balkanism. The key difference is that Balkanism is linked to a transitory and liminal space, while balkanization is used as a set of processes to remake space and organize knowledge for purposes of Westernization and in the name of eradicating the Other.

The construct is also encompassing of a set of values and norms; an ordering one needs to comply with if one is to attain the civilized 'normality' and stability found in the Anglo-American and Western world. According to James Der Derian, such division disregards the ongoing history of balkanization from the Roman Empire until the present.[12] What is more, he notes, when balkanization is

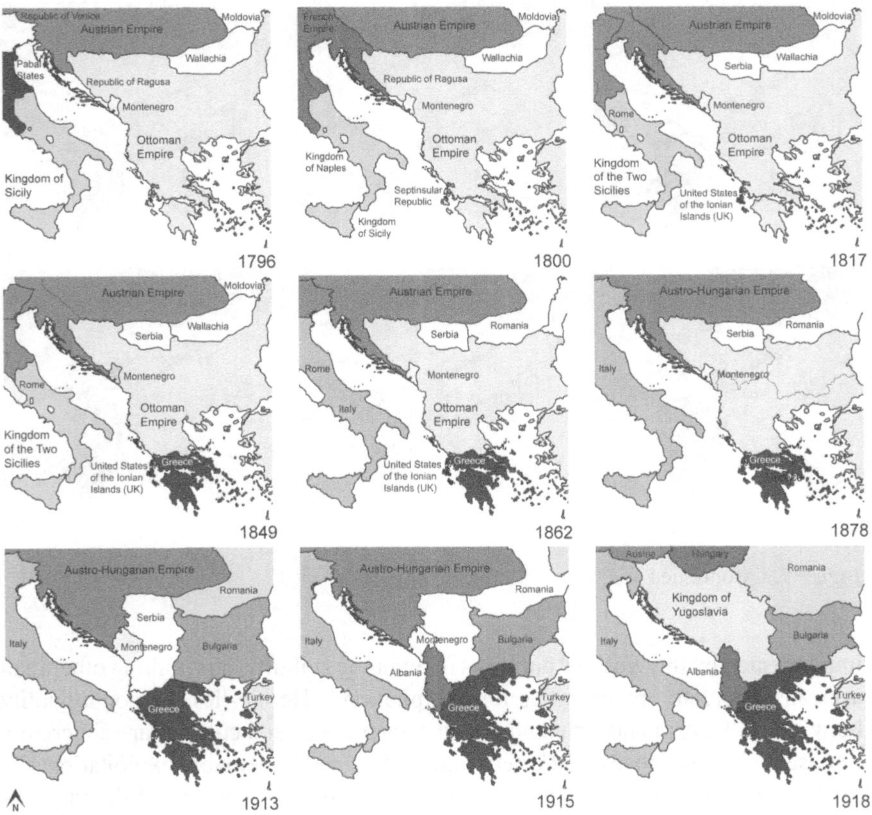

Figure 1.1 Balkanization of the Balkans from the 18th Century to now.

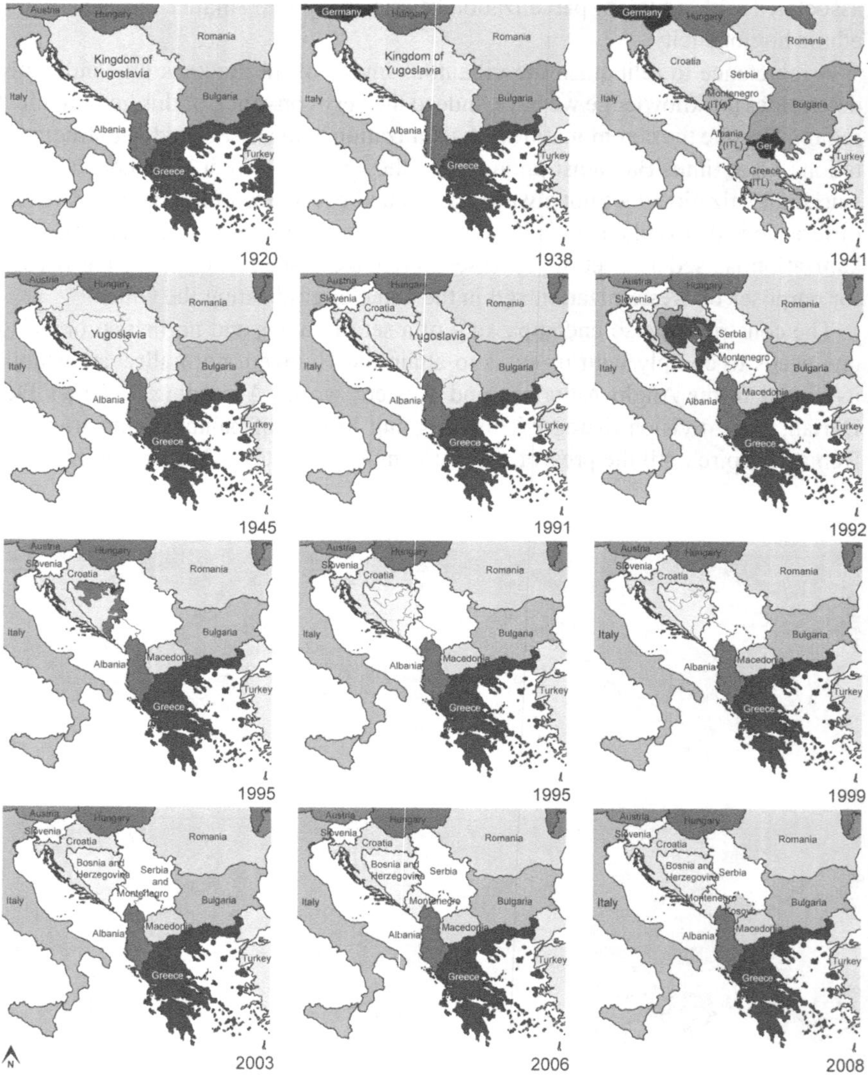

Figure 1.1 Continued

implemented by the West for purposes of attaining stability, it is nothing other than a means to create opportunities for exploitation.[13] Herein lies the commonality between Balkanism and Orientalism: whenever zones and cultures are concretely represented for purposes of Western knowledge, and undeniably exploitation, the motives are geared towards imposing Western control.[14] However, while fragmentation of settlements, cities and territories has a long historical trajectory, often pursued for accretion of settlements, balkanization as a process is associated with the Balkans. It is because this region is identified as an in-between zone, a region of paradoxes and friction. The implementation of balkanization is more fervent as it is contingent on the existence of this peripheral zone of excess as buffer between the Eastern Orient and the Western Occident.

In the contemporary world, the term balkanization[15] is specific not only to the Balkans and its circumstances. It has also been used to describe, for example, the impact on the US of immigration on its urban planning, legal system, medical practice and the internet.[16] These more recent connotations of balkanization tend to erase the derogatory associations with balkanization as undertaken in the Balkans, the breaking up of a large and violent geopolitical zone into smaller states for purposes of bringing peace and order. However, the contemporary non-geopolitical usages still retain the power-oriented associations of an underlying lack of dialogue. To understand the complexity of balkanization, and its relation to Balkanism, there is a need to frame it in terms of the former Yugoslav context.

Yugoslavia, Balkanism and balkanization

The 'truth' about the discourse of Balkanism is to be analysed from the periphery. The part of the periphery that is analyzed is the one that has posed the most alternative way of deploying balkanization – the former Yugoslav context. Against the actual fragmentation of the old Ottoman Empire into the other Balkan countries (Albania, Bulgaria, Romania, Greece and Turkey), the formation of Yugoslavia (land of the southern Slavs) in 1918 approached balkanization in reverse, by grouping together territories where different South Slav peoples lived – meaning that the entity created was heterogeneous. In other words, while some areas of the Balkans were fragmented for purposes of creating an ethnically pure nation state model (a homogeneous structure superimposed onto previously heterogeneous nations), Yugoslavia not only retained but also extended its process of heterogeneity. This is the significance of the former Yugoslav context, and the reason behind this book's focus on the region.

With the 1999 Operation Allied Force being the 20th Century's last military intervention, the significance of studying this intervention in terms of its spatial and Balkanist implications is paramount. This is because both the intervention and its timing resonate with the complexity of deliberate balkanization and often disguised violence inflicted. The deployment was not only on the various aspects of a city, but also implemented by 'obvious' mechanisms such as the military's use of technology, or less obvious ones such as the economy, law, mass media and colonization. The significance of examining the intersection of Balkanism, balkanization and the intervention is also because this military operation has not been given enough attention, especially from a spatial and architectural perspective, despite the academic interest in and investigation of the discourse of Balkanism that coincided with the 1990s disintegration of the Socialist Federal Republic of Yugoslavia (SFRY).[17] Balkanization was revived with the dissolution of the former Yugoslavia.

It must be noted that while this volume predominantly frames Balkanism in terms of the zone and peoples of former Yugoslavia, the Balkanism discourse is broader and encompasses all the areas of the Balkans. The analysis of Balkanism here is undertaken via Yugoslavia not only because the zone has implemented balkanization in ways different from the norm, but also because the 1990s secessions from Yugoslavia was often framed as the Balkan war; the 'powder keg' of Balkan violence and balkanization was epitomized in Yugoslavia. In other words, the reverse balkanization that was implemented in Yugoslavia for purposes of achieving heterogeneity was largely crushed by the 1990s events that occurred within that territory.[18] The familiar characterization of the Balkans as a savage and bloodthirsty place was not only resurrected during those events but also used to affix and determine this space.

The disintegration of Yugoslavia started with the relatively peaceful secession of Slovenia in 1991 and continued through more violent conflicts in Croatia (1991–95), Bosnia and Herzegovina (1992–95), Kosovo (1998–99) and Serbia (1999).[19] The focus here will be on these latter conflicts, as the ones most resonant with violence during the period of the 1990s. The entity's heterogeneity was crushed in a parcelizing balkanization driven, as reported around the world, by elements of ethnicity and nationalism. With a historical imaginary as a permanent source of ethnic hatreds and nationalism, Yugoslavia was bound to erupt in violence; after all, its historical 'truth' is written in blood. Not only was the dissolution 'surely' a yearning for democracy and freedom, but also it proved that Tito's espousal of non-alignment and the economic merging of socialism and capitalism had reached its expiry date with the democratic revolutions of 1989 in Russia and Eastern Europe.[20]

The apparent self-evidence of this imaginary was demonstrated with numerous books, testimonies, memoirs and diaries about Yugoslavia and the 1990s violence; expert punditry emerged overnight and was largely accepted without question.[21] Western positions were shaped simplistically rather than through an effort to understand Yugoslav affairs in their complexity.[22] The mass media largely abandoned its critical brief when reporting on the dissolution. Balkanization of Yugoslavia seemed to be a given; it was, after all, a zone of historical instability considering its geographic position between the East and the West, a region of disputes and backwardness where democracy was foreign in both concept and practice. However, the 1990s dissolution was a very different invoking of balkanization to Yugoslavia's inceptionary processes of coexistence and diversity at the start of the 20th Century.[23] Indeed, it was its reversal.

Critically analysing the Yugoslav context will assist in framing the many ways *balkanization* as an *act*, a *method* and a *set of processes* can be deployed: from its opportunistic application to the analysis of balkanization in terms of more recent enactments since the 1990s. Particular attention will be paid to the ways in which this analysis may aid in understanding the more coercive implementation of balkanization that is globally taking place today. This is principally important because the spatialization of violence has become entangled with economic, mass media, technological and legal apparatuses; indeed, these entanglements make it harder to demonstrate evidence of violence.

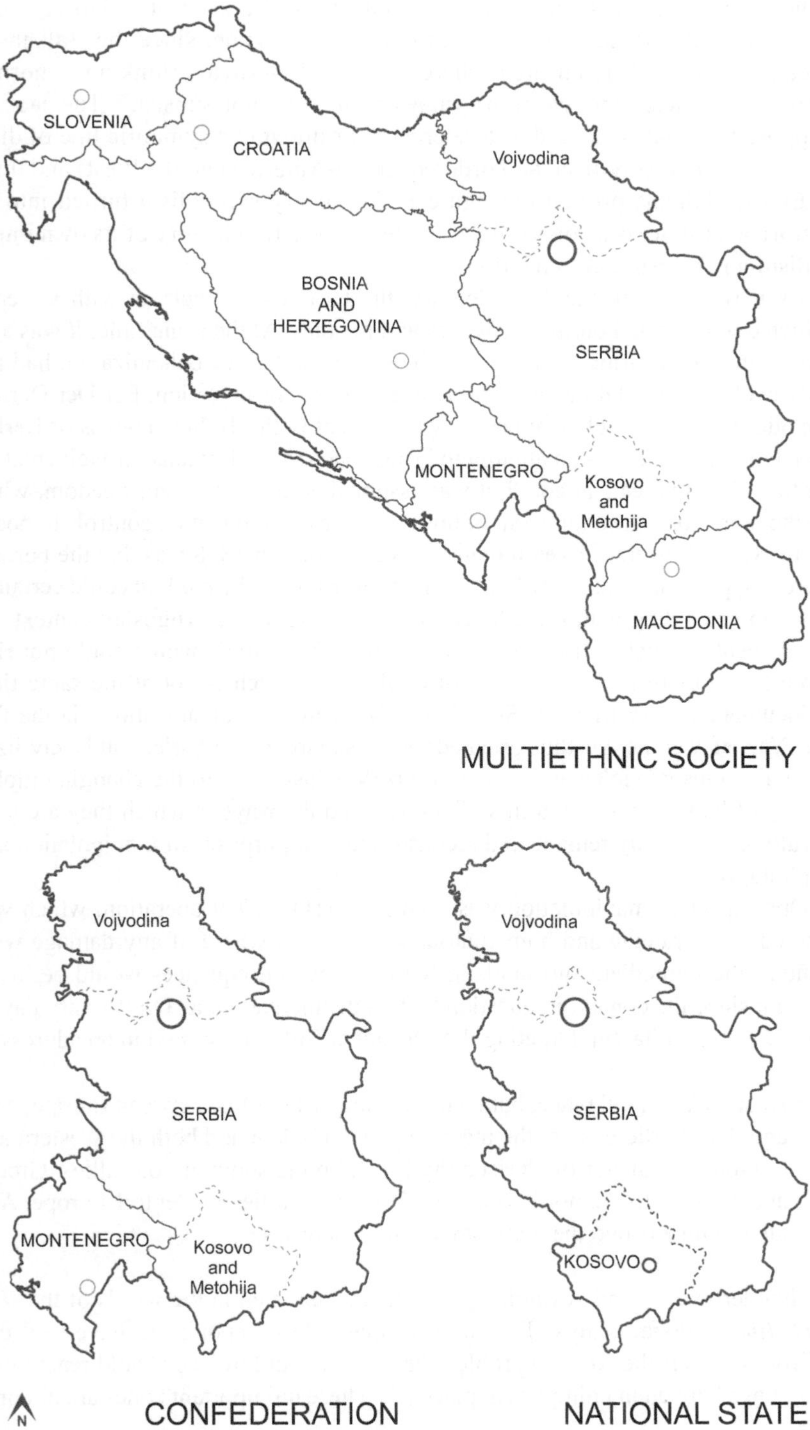

Figure 1.2 Balkanization of Yugoslavia in the 20th and 21st Centuries.

When balkanization is brought into being in the Balkans, it is largely done to override the values and practices found in this zone, since the Balkans is broadly perceived as an area where – using Todorova's thinking – normal (Anglo-American and Western) values simply do not subsist.[24] The lack of supposed normality is used as a reference for drawing a symbolic line of division between us (civilized and ordered Anglo-America and the West) and them (barbaric Balkans prone to slaughter). The binary symbolism turned into an historical artefact is an opportunity to cleanse Western history of its own imperialism and barbaric colonizations.

As it was used in the 19th Century, the term was associated with violence inflicted upon neighbouring territories and peoples. At the same time, it was also the means for securing peace. Thus, the understanding of balkanization had not only cyclical connotations but also inherent scope for inversion. For Der Derian, the oddity of utilizing balkanization was evident in the 1878 Congress of Berlin, which on one hand was an attempt to bring peace to the Balkans through creating artificial boundaries that could, it was assumed, secure unity and freedom, while on the other hand the attainment of this peace was contingent on controls imposed by the Great Powers.[25] Even though the assumption in 1878 was that the borders were mapped onto ethnically homogeneous territories, the method could certainly have no grounding within the heterogeneity of the former Yugoslav context. As a comment on such self-contradiction, Walter Benjamin's words could not ring more true: "there is no document of civilization which is not at the same time a document of barbarism."[26] Studying Balkanism and balkanization via the former Yugoslav context is thus oriented towards mapping barbaric, that is, civilized spatializations of violence. As such, it proffers insights into the changing global modes of balkanization in terms of control, and the ways in which they are temporally and spatially remade and reconstructed for purposes of manipulation and exploitation.

One aspect of manipulation was seen in NATO's 1999 operation, which was framed as a strategic and humanitarian intervention where, if any damage were done to the immediate surrounds in Kosovo, the consequences would be insignificant since the context in question had no distinctive value. On the very day of NATO's 'humanitarian' targeting, US President Bill Clinton said in an address:

> Kosovo is a small place, but it sits on a major fault line between Europe, Asia and the Middle East, at the meeting place of Islam and both the Western and Orthodox branches of Christianity [. . .]. To the south are our allies, Greece and Turkey; to the north, our new democratic allies in Central Europe. And all around Kosovo there are other small countries.[27]

Such a derogatory and devaluing position was reflected in the words of the *New York Times*' global analyst Thomas Friedman. He is reported to have said that "[f]rom the start the Kosovo problem has been about how we should react when bad things happen in unimportant places."[28] These unimportant zones are not only areas that cannot be located on a map or where security is a potential concern considering the historical clash of Western and Eastern civilizations and religions. They are also, and more importantly, zones that evade straightforward distinctions, the difficulty of representation being due to their liminal and *under construction* identity.

The tying of rhetoric to the need to eliminate complex liminality – and ultimately difference, considering the historical presence of heterogeneous cultures, values and ethnicities – is a significant form of violence, since it is allied to the belief in and need to subscribe to a set norm and identification, which is always measured in terms of Western European and Anglo-American ideals. According to Noam Chomsky, Kosovo was the prime example used in an argument that, for the first time in history, "states were observing 'principles and values' under the guidance of their 'noble' and 'altruistic' Anglo-American tutors, and that the UN Charter must be revised to allow the West to carry out 'humanitarian intervention'."[29] The relationship between the intervention and balkanization is that it was used for purposes of first fragmenting the territory and then destroying architecture and infrastructure as well as affecting the urban and architectural spaces of everyday life. The attack was aimed towards subordinating and then immobilizing various lines of communication such as transport routes, media networks and the alternative economy.

Balkanism and balkanization are complex. Thus, this volume will broadly draw together a range of currently quite disparate discussions on balkanization from a variety of fields (urban planning, sociology and anthropology, internet, law and medicine), to the more specific focus to do with the former Yugoslav context, in order to address the growing significance of the term balkanization and its implications. The lessons learned from an in-depth analysis of the Yugoslav context can be used to examine other global contexts and cities. This is particularly important because little critically detailed and synthesized research has been done in respect to the phenomenon, and virtually none from the architectural and urban perspective.[30]

Spatial role of politics and balkanization

Architecture and balkanization

The necessity to analyze balkanization from an architectural and urban perspective is because it reveals political and, often veiled, aspects of balkanization. Urbanism and architecture do not exist in isolation from social history and politics of balkanization, and as such tend to become a target, as they signify the Other. To target and remake cities and architecture today is not only to fragment, hollow out and reconstruct, but also to understand the consequences of how economics, law, mass media, technology and the military (the global politics of control) intersect with architecture. Control of the Other is possible to the extent that the forces of control intersect with architecture. Balkanization is used not only to fragment and unmake the Other but also to remake it, since the process of unmaking affords the opportunity to remake and instate the preferred (Western European and Anglo-American) ways of living which, in turn, assist in managing the Other.

Architecture and the urban are political,[31] and this volume will address how the noted forces of control play out pragmatically as well as in terms of identity, ideology and symbolism. Urbanism and architecture do not exist in isolation from social history and politics. Examples of architecture and the urban from the former Yugoslav context and, in particular, Belgrade and the 1999 NATO intervention on Serbia, will be examined and used as a lens through which to expose and question the hegemonic implementation of balkanization. This is for two key reasons. The first, related to the broader Yugoslav context, is that it shows how a country that was a key leader in the Non-Aligned Movement and that instated an alternative thinking about architecture, socioeconomics and politics, has been balkanized for purposes of instating Western power and control. The second reason, specifically related to Belgrade and Serbia, is that the 1999 military intervention offers the first insight into the changing modes of balkanization into the 21st Century. The insights that arise may help in thinking how balkanization is present in cities outside the regions mentioned, despite these not being addressed here. To address the relationship of balkanization to broader global urbanization would require writing a second volume; however, some of the ways in which balkanization and politics of architecture may be considered are identified ahead.

From a pragmatic perspective, architecture is the very agent and medium of violence, and associated with understanding the processes by which balkanization has been implemented. Speed is essential as a way to control, survey, map, and introduce consumerism and mass privatization as a preferred economic context. In this process, the law is deployed to enforce ad-hoc campaigns and transform values that in the long-term suit the agenda of market economics and corporate capitalism. The effects of unmaking and remaking facilitate control. Control is not always completely apparent, and is brought into being through national, EU and international policies to suit a particular hegemonic agenda to do with security.

The ideological and symbolic aspects of architecture are related, and are to do with aesthetic and verbal rhetoric, including the ways in which it is presented and mobilized through architecture for political and economic gain. Rhetoric is performative and exercised for purposes of historical myth making and attaining supposed security. Balkanization becomes the preferred means of achieving these aims. This agenda is politicized spatially by utilizing the law, language and mass media, and connected to an imagery of brutality; all of which help flatten the complexity of place and history, whilst also exposing the danger that arises when spectacle infiltrates politics.

Political matters to do with identity of architecture and the urban include the attempt to eliminate alternative positions seen in everyday life, values and rights, as found in the discourse of Balkanism. Ultimately, balkanization is enacted with the aim to negate difference, an attempt to achieve ethnic purity either by proclaiming a sovereign right to land or through the construct of that territory as lacking civil values, thus establishing the need to flatten it. Negating difference is geared towards immizeration and dehumanization, and architecture is used as a significant way of spatializing the agenda. Often, the urban and the architectural are destroyed to the point that inhabitation becomes impossible.

Some attempts to address the process of balkanization as a positive strategy and spatiopolitical facilitator of "bottom-up" initiatives against the "top-down" power structures are seen in the thinking of Srdjan Jovanović Weiss.[32] For Weiss, fragmentation ultimately leads to diversification, as seen with the territorial fragmentation of the SFRY in the 1990s. The official Serbo-Croatian language was broken up to identify four languages (Bosnian, Croatian, Montenegrin and Serbian).[33] In other words, this change has brought on "multiplication." Another positive, according to Weiss, was seen in the branch of Deutsche Bank in Manhattan where balkanization first became a design strategy to fragment the 40-storey monolithic structure so that the interior becomes completely hollow, and second to use the fragmented self-similar entities for purposes of reassembling them in a different way.[34] Weiss' thinking is that balkanization is a strategy and as such is to be implemented for purposes of "learning with" the Other; however, the Other is "accepted as missing."[35]

This volume does not follow Weiss' thinking in attempting to eliminate the liminal Other, since balkanization ultimately is not about self-identity but about control. The Other not only exists but also is a key aspect to rethink questions of norms, urban conditions and everyday living. Also, thinking diversification, including the identification of agency as "bottom-up" or "top-down" is much less straightforward considering the complex affair of multiple agencies enfolded in spatializing power, discipline and violence – including the networks of control societies – that are operational today. The convolution of the modes of control and the conditioning of norms are more carefully and surreptitiously regimented in control societies since, utilizing Gilles Deleuze's thinking, "snake's coils are even more intricate than a mole's burrow."[36] These are certainly present in the remaking of former Yugoslavia.

Spatial violence and spatial remaking

The fragmentation of Yugoslavia in the 1990s and its simultaneous reorganization were supported by the Western-devised Dayton Agreement, which made the assumption that peace can be administratively instated by establishing homogeneous ethnic enclaves. For a region known for its ethnic heterogeneity and mixing, this is not only a complete violation and misunderstanding of culture and history but also in itself the epitome of violence. By removing the geopolitical association of the Balkans and replacing it with a purely political one, the integration of the Balkans with Europe and the West becomes contingent on restricting diversity to avoid difficulties that supposedly arise when negotiating multiple ethnic identities and values in one territory. What this thinking shows is short-sightedness when forming states on the basis that peace and order cannot be achieved without ethnic separation. The Western preference for sovereignty in terms of a common ethnic identity is an instatement of colonialism: newly independent and war-ravaged countries established on this basis not only have become even more accessible to foreign organizations and investors, but also their reconstruction and existence are contingent upon entering into debt-inducing relationships with such entities.

Society in the former Yugoslavia was regulated, striated and overlaid with surveillance. The mode of this regulation, this striation, was different from that practised in the EU and the West. Part of the reformation after NATO's 1999 intervention has been oriented towards reterritorializing and incorporating this differently striated space into the wider European and global network for the purposes of control, surveillance and defence. The economic loans, largely provided by the World Bank, the International Monetary Fund (IMF) and similar donors, are a seamless way to militarize the society and urban life; NATO and the economic donors are an entangled monopoly of global politics enacted in the name of stabilization.

Graham accurately observes that "[t]he rich cities of the advanced capitalist world profit from urbicidal violence, which deliberately targets the city geographies of the Global South to sustain capital accumulation."[37] Provisionally, supposed stabilization is dependent on a two-step process; first, creating a system of exclusion by identifying an insignificant Other zone as problematic, and second, asserting the need for it to be politically altered. One instance of this is seen in the addition of *Western* to the name Balkans. The name change is a way of eliminating the right to difference considering that the organizations and global forces involved in loan financing states, such as contractual economic impositions of the EU and the World Bank, also indirectly control the ways in which urban areas are developed and expanded.

What makes the control more seamless is that the reconstruction of destroyed Yugoslav cities and institutions is tagged with a post-socialist and transitional agenda. The transitional tag is used as a way to excuse a period of repression and backwardness (pre-1990s Yugoslavia) as preceding imminent democratization and progress following the termination of the conflict. Thus, in order to address balkanization as it applied to the violence of cities in the former Yugoslav context, and to map these examples in terms of national and international perspectives, it is necessary to consider the very categorization of this region as undergoing transition. In other words, the use of the transitional tag is an opportunity to assert a dependency relationship and supervision by foreign investors; to be democratizing is to be in the regulatory hands of private global corporations.

That the dimension of geoeconomics and politics is entangled in monopolization and reduction of complexity was seen with NATO's 'humanitarian' intervention. The association of the 20th Century with violence and inhumane warfare has spurred global politics to seek means to enforce justice 'humanely' in the 21st Century. The humanitarian intervention of 1999 was certainly not an exercise in philanthropy and was not mutually beneficial. Hybridizing humanitarianism with military warfare hijacks and colonizes the present while attempting to prevent the future from occurring. Today's power is exerted in an attempt to eliminate the possibility not only to engage with the politics of history, but also to activate it.

The activation is dependent not only on removing the overarching understanding of *a history* but also on thinking spaces beyond those external to the privileged norm. Similarly, to activate history is to think outside and beyond the networks of control.

In terms of the Western Balkans, control has been achieved by constructing history along the lines of relatively simplistic Balkanist imaginaries and denying this region an opportunity to exist as a liminal zone. The world order of today is politically just only if it assumes deregulation of alternative ways of thinking socioeconomics and welfare through processes of balkanization; it is bureaucratization made possible through control implemented in 'humane' ways. According to Chomsky, the 1999 intervention heralds the arrival of a use-by date for international law and for NATO's ability to orchestrate the definitions of justice.[38] This is particularly significant in terms of understanding more contemporary iterations of balkanization, and the reason why this volume dedicates considerable space to its analysis, predominantly by focusing on the city of Belgrade and NATO's 1999 Operation Allied Force more generally.

The 'humanitarian' agenda is seemingly at odds with Leon Trotsky's thinking on the Balkan Wars of the 1910s, namely that war always reveals our barbaric ways even as we believe that progressiveness and civilization – all that is admirable in material acquisitions, customs and habits – foreground our society.[39] With the 1999 intervention, the barbaric side of NATO's 'humanitarian' agenda has been somewhat concealed. This apparent contradiction is deceptive, made possible precisely because the threshold of conventional understanding of military violence on one hand, and compassion on the other, being blurred. Unlike the compassionate side of humanitarian aid directed towards all peoples in need, military humanitarianism creates a clear line between victims and perpetrators on the basis of ethnicity. What has lent plausibility to the possibility of a 'humanitarian' agenda in military interventions is the fact that the interventions are carried out from a distance, through smart bombing and remote control on computer screens. NATO's 1999 Operation Allied Force was the first war prosecuted from the air alone.

Technology has been used not only to obliterate the limits and implications of violence but also to disable the possibility of engagement with the paradoxes of humane warfare now that it is transmitted to a global audience via the televised screen of spectacle. To spectacularize violence is also to neutralize it by aligning it more closely with the conventional understanding of humanitarian aid. Yet the ability to believe that military interventions can be humanitarian and an attempt to bring peace and democracy reflects the reality that we live in a society that is not only controlled but also sedated. The effect of targeting and framing zones from a distance thus assists in eclipsing the complexity of violence as well as eliminates the possibility to believe that any worthwhile life exists in the Western Balkans. After all, Clinton identified regions surrounding Kosovo as simply small countries unworthy of the name. Seemingly, to engage with the implications of humanitarianism in the former Yugoslav region of the Western Balkans is balkanizing politics itself.

Global politics is not a means to eliminate borders and enable a cosmopolitan culture of inclusivity and solidarity.[40] It is instead somewhat resonant with Manuel Castells' "network society" where governing is no longer institution-bound but spread globally, meaning that power is found in new networks of information and images that are, perhaps, less rigid.[41] The contemporary examples of global politics certainly resonate with Zygmunt Bauman's "negative globalization," where current examples to do with coercion and instability facilitate the opportunity to also ask questions to do with control and common privileges.[42] Likewise, it may be argued that the fully fledged expansion of global politics which took hold with the fall of the Berlin Wall, the Velvet Revolution and the collapse of Cold War international politics has meant that the military has not only embraced but also drives global and largely privatized news media, market economics and securitization for purposes of exercising control over any system still keen to resist the Anglo-American and Western one. In terms of Belgrade and Serbia after 1999, it is a low-intensity and high-tech ground conflict deployed with the intention of achieving desocialization through reformation of urban spaces and society. It is being facilitated by rapid changes to infrastructure, law, privatization of companies and transformation of territorial relationships. The intent is geared towards transforming a once-socialist country under Titoism – an unconventional socialist model which had an alternative approach to welfare, economics and the law – to a Western one under Western colonial hegemony.

More so, it is not just a matter of militarization being intertwined with these dynamics, but that technology – that is, speed – has been used to accelerate and extend the limits of conventional war strategies into urban infrastructure. Contemplating the modern world of militarization, Paul Virilio observes that it

> is a world in motion, expressed in translations of strategic space into logistical time, and back again. It is a history of cities, partitions, trading circuits, satellites, and software; of a political landscape governed by competing technologies of surveillance, mobilization, fortification.[43]

Virilio suggests that the current society is distinguished by speed and militarism rather than capitalism, since the accumulation of capital would tend to stop the acceleration of technological advancements.[44] However, the argument to be put forward here is that the balkanization of Yugoslavia in the 1990s demonstrates a different scenario. By analyzing the ways in which balkanization has been carried out in the former Yugoslav context, this volume will argue that the drivers are both speed – that is, militarism, including the extension of military control into every facet of society – and capital, in terms of the proliferation of corporate capitalism, consumer culture and debt. In other words, militarism and capital coexist; each contains within itself elements of the other. Militarization is the means to facilitate and stimulate the speed that consumerism and debt manifest.

Considering that speed implemented is *also* a measure to prevent violence, as seen in the humanitarian launching of missiles in 1999, the opportunity to engage with the complexity of a technological culture and consider its implications from different perspectives is undermined. To have an understanding of a technological culture of speed would mean that one would start eroding the current popular disengagement from politics, since the rhetoric and imposition of security measures have largely become the driver behind every state and global activity as well as their political legitimization. It is inevitable that, with the privatization of information and data, the understanding of the military and conflict also becomes privatized and interest-driven; the general rhetoric that is transmitted to the public is largely uniform and synonymous with flattening and neutralizing information through appropriation. A constructed story on a television screen can be accepted as accurate in circumstances where certain violent events are readily taken as evidence and confirmation that a certain nation is 'naturally' prone to acts of violence. Violence on the sports field is one example. Virilio has noted that "football simulates primordial territorial clan warfare, and the supporters of Red Star Belgrade are quickly recast [by the West] as the shock troops of Serbia sweeping through Bosnia."[45]

The tendency to regard certain spaces and histories, such as that of the Balkans, as prone to violence not only treats identity as fixed but also aids in discursively positioning such supposed zones of violence and threat as responsive only to the use of force, as any other approach will not be comprehended. Seemingly, balkanization – as it is conventionally implemented for purposes of territorial and ethnic separation – is a process of not only flattening the complexity of violence but also legitimizing these global networks of control. The enactment and legitimization of balkanization is made possible through the invocation and implementation of security. Any attempt to challenge the status quo is interpreted as a clear sign of refusal to enter the civilized world, as was seen during the non-negotiability of the legal conditions in NATO's Rambouillet Agreement.[46] Security is an important element of this narrative, in terms of negotiating disparities of cultural coexistence, nationalism and ethnic hatreds between Serbs and Albanians in Kosovo; intervention to impose international security measures was presented as a prerequisite for peace and stability as well as an obligation to respond to acts of violence.

The force of balkanization

While balkanization in its conception is not detached from the Balkans and the discourse of Balkanism, the multidisciplinary use of balkanization means that it has a fluidity and force that cannot be bound geographically. At the same time, the reason why there is a possibility to think about the discourse of Balkanism having potential is precisely because it not only embeds the open-ended complexity of excess and abnormality but also because it operates in ways that build paradoxes and alternatives into the networks of control. This book approaches excess and fluidity as positive, but not in a way where decentralization and fragmentation are seen as a dialectic bottom-up process, or a top-down imposition. Instead, by utilizing the very thinking and spatial practices found in the balkanization of the former Yugoslavia, which has historically and geographically been seen as a frictional zone, this volume uses the SFRY as a case study; a means to challenge the current writing on both the Balkans and balkanization. In other words, when Western deployment of global politics and violence come into contact with a space that is operating with a different flexibility and logic (such as Balkanism), alternative understandings of violence engendered with the disintegration of Yugoslavia may be projected. The alternative queries that arise from addressing the balkanization of the former Yugoslavia are inclusive of, but not limited to, at least three significant modes of rethinking violence. The first rethinking is to do with the application of the label ethnonationalist conflict to war and violence that occurs in Other countries, thereby shutting down the possibility of exploring violence in Western cities that itself may derive from conflict among multiple ethnicities.[47]

The second query to do with violence involves the way that potential to consider the role external forces (the West) have played in facilitating the destruction of cities in the former SFRY has been eliminated. Or, that since the 1999 NATO incursion, some former SFRY Republics (now independent countries) have been granted access into the EU – such as Croatia – thus making possible the erasure of certain forms of violence from the immediate and medium distanced past (SFRY dissolution to WWII). Granting the possibility to efface the record of acts of violence conditions how particular spaces and histories are portrayed in politics and the media. (In)directly, the conditioned political envelope accommodates other Balkanist zones and peoples to be sided with the imagery of savagery and through the lens of perpetrators.

The third disguised mode of violence is even more complex. The query is why act upon violence in Kosovo in 1999 rather than considerably earlier, considering the centuries-long violence in this territory, a history that encompasses the period of Serbia's loss of the territory of Kosovo to the Ottoman Empire despite the self-celebrated bravery of Serbs when defending and then losing the 'cradle' of Serbia's homeland. In other words, what was at stake in 1999 that made this NATO Operation unavoidable? Addressing this via the 1999 targeting is paramount, particularly the targeting of Belgrade as capital of what was then known as the Federal Republic of Yugoslavia (constituted of Serbia and Montenegro). The implications of this will extend the agenda and impact of balkanization and Balkanism by addressing, through specific case studies, how architecture and urbanism intersect with matters to do with economics, warfare, environment, and social and individual values.

Thus, the exploration of Balkanism and balkanization will be used as a critical tool to analyze the contradictions as well as implications of control and violence on architectural and urban levels, including their implementation and consequences when it comes to standards of living (economic, social and individual). Having said this, the focus is not on showing society as it really is; to believe in such a possibility is to eliminate the need to imagine. Instead, the focus is on putting forward questions, placing sparks, stirring frictions and proffering alternative positions to the general assumptions about Balkanism. The critical agenda thus addresses alternatives through the very reinterpretation of the discourse historically and spatially, which indirectly disturbs and destabilizes historical 'truths' to do with Balkanism and balkanization. The disturbance enables a different understanding of history by opening up ways of thinking and spatializing Balkanism and balkanization that have largely remained closed. Neither does this analysis react against the current political structures. To have a reactionary position does no more than institute a cyclical active-reactive relationship, which not only strips away creativity but also fails to recognize that alternatives exist. To react is not to affirm life. The study is instead actionary; the outside – utilizing Deleuze's thinking – does exist.

The volume will be framed from a Foucauldian angle whereby knowledge will be generated from the perspective that it is not universally given; it has simply been formalized as acceptable and universally true.[48] In other words, to loosen the supposed truth about the discourse of Balkanism and to critique the relations that have facilitated this knowledge. Even so, the volume does not just put forward critique but also offers alternatives; the alternatives are affirmative, despite attempts to negate them.

In order to address the framework outlined, the book will be broken into six sections. Chapter 2 *Balkanism and balkanization: fragmentation, grouping and excess* will extend the preliminary framings on Balkanism found in this introduction. The focus will be on disassembling and critically analyzing the formation, alteration and implementation of Balkanism and balkanization in a number of different fields. This will be done not only from a historical perspective in terms of addressing questions to do with territory, ideology and politics, but also to examine the ways in which this term has been deployed in fields as wide-ranging as urban planning, sociology and anthropology, international law and the digital world. Against this broad and global analysis, specific focus will be directed towards addressing how balkanization was pursued as an initiative within former Yugoslavia in the late 19th Century and the bulk of the 20th Century, deployed as it was in radically novel and creative ways (grouping through diversification, solidarity and individual rights). Similarly, the more recent utilization of balkanization within this context mimics the more commonly understood premise of balkanization as being fragmentation and parcelization for purposes of establishing majority ethnic enclaves. The analysis of balkanization through both, a broad and global lens, and a focused and regional one, will assist in extrapolating the complexity of balkanization as well as in mapping the shifts in its implementation and understanding over time and in different fields. The focused mapping will address the implications of Balkanism and balkanization across film, public space and architecture.

The third chapter, *Yugoslavia in balkanization: beyond civil war, beyond ethnonationalism* explores how fragmentation that occurred in the SFRY in the 1990s plays a key role in understanding the more coercive implications of balkanization taking place today (within this region, and globally). The argument will be that the Yugoslav context was a zone where these coercive acts were simultaneously entrenched and tested for subsequent spatial deployments of balkanization globally. Against this proposition, Yugoslavia's disintegration will be framed against the ideological and political agendas according to which the country was constructed and later balkanized. In other words, the historical framework will contribute towards the understanding of fragmentation in respect to violence (of both obvious and less obvious kinds). It will go beyond the proposition that the dissolution of SFRY was purely an example of ethnonationalism or that it was a civil war. The balkanization of SFRY was a complex enactment of spatial violence, a complexity first noted during the country's disintegration in the 1990s as fragmentation was related to war, international law, international security enacted in the form of UN troops and media, and the fragmentation of the ethnically diverse country. The discussion will be contextualized by examining memory and destruction, and the appropriation of both of these when it comes to the reconstruction of cities such as Vukovar and Dubrovnik in Croatia, and Sarajevo and Mostar in Bosnia and Herzegovina (the cities that have received the most mass media and academic attention in response to the 1990–95 disintegration of the SFRY).

The aim of the fourth chapter, *Allied operations: present-future partnerships of humanitarianism, peace and victory* is to explore the more coercive aspects of balkanization in terms of spatial violence. This considers, in particular, the ways in which the first humanitarian air intervention – 1999 Operation Allied Force – facilitates exploration of the methods that shape these new modes of balkanization: from history and memory to control via mass media. The virtuality of violence has very different connotations for those viewing it as presented through spectacular imagery than it does for those experiencing the effects of 'humanitarianism.' To give context to the discussion, the review will be focused on specific examples of architecture, monuments and landscape zones in Belgrade in terms of identity and reconstruction.

Circulating violence: decomposition, dispossession and control, the fifth chapter, will analyze more closely not only the relation between Balkanism and balkanization but also the geopolitical implications in terms of the more hidden enactments of violence during Operation Allied Force – the logistics of the Operation and the effects on the environment and human bodies, not least from the depleted uranium contained in NATO's high-tech projectiles. The aspect of spatialization will be addressed through the circulating forces of violence to do with media and high-tech infrastructure such as satellites and unmanned drones as well as the legal procedures themselves that facilitated the Operation. The analysis of resources and infrastructure in Serbia and Kosovo, as another instantiation of violence spatialized in the aftermath of 1999, will critically address the relations between destruction and reintegration in terms of balkanization.

Chapter 6, *Territorializations: transitions, thresholds and transgressions* will address the means and methods of, as well develop awareness to do with, current and future (altered) agendas of fragmentation in terms of gentrification, violence and control taking place along the margins of Belgrade's waterfronts, and the impact of the same when thinking about marginal groups such as the Romani and the economically less privileged. As such, this chapter provides an exploratory analysis when thinking about these more coercive and softer enactments of balkanization. These softer spatializations of balkanization and, in effect, violence, will be mapped against urban planning and infrastructure as well as the role that public spaces play in this reterritorialization of Belgrade. In particular, how the transitional tag is used as means to instate particular thinking and methods to do with urban planning (associated with a tightening of operational forces of control) and ways of deploying Balkanism to think of alternatives.

The last chapter – *Present spaces, present times: other spaces, other times* – will draw together the key ideas raised in the previous chapters but, more importantly, focus on balkanization in terms of its variable liminality, since the very concept has an embedded mobility. Any fixity of the concept – or any attempt to clearly define it – is secondary to the flexibility and mobility found in its interpretation. Thus, mobility will be approached as productive and as a creative potential. This chapter will reconceive balkanization by amplifying its very heterogeneous potential (diversity seen for a large part of the 20th Century in Yugoslavia, and the Balkans prior to the mid-19th Century) in order to open up global thinking about questions to do with the economics of fear in terms of territory, security and displacement. More broadly, this will address how the more progressive thinking about balkanization (associated with diversification and constituency) may be deployed to start the process of rethinking human rights, values and humanitarianism.

Notes

1 Maria Todorova, *Imagining the Balkans* (New York: Oxford University Press, 2009), 3.

2 Dušan I. Bjelić, "Introduction: Blowing Up the 'Bridge,'" in *Balkan as Metaphor*, ed. Dušan I. Bjelić and Obrad Savić (Cambridge, MA: MIT Press, 2002), 7.

3 Bjelić, "Introduction: Blowing Up the 'Bridge,'" 7.

4 Edward W. Said, *Orientalism* (London: Penguin Books, 2003).

5 Katherine E. Fleming, "Orientalism, the Balkans, and Balkan Historiography," *The American Historical Review* 105:4 (2000): 1218–1233.

6 Alexander Kiossev, "The Dark Intimacy: Maps, Identities, Acts of Identification," in *Balkan as Metaphor*, ed. Dušan I. Bjelić and Obrad Savić (Cambridge, MA: MIT Press, 2002), 165–191.

7 Todorova, *Imagining the Balkans*, 16.

8 The change of name occurred during the European Council meeting in Vienna on 11–12 December 1998, only three months before NATO's 24 March 1999 targeting. The initiative occurred only after Romania and Bulgaria started the open-membership dialogue with the EU. Notably, for the EU, the Western Balkans comprise Croatia, Bosnia-Herzegovina, Serbia, Montenegro, Macedonia and Albania.

9 Stephen Graham and Lucy Hewitt, "Getting off the Ground: On the Politics of Urban Verticality," *Progress in Human Geography* 37:1 (2012): 74.

10 Characters such as č, ć, đ, š, č, ž are only used where authors spelled their names using these letters in the original source. In all other instances, letters that appear in the English alphabet are only utilized.

11 Rastko Močnik, "The Balkans as an Element in Ideological Mechanisms," in *Balkan as Metaphor*, ed. Dušan I. Bjelić and Obrad Savić (Cambridge, MA: MIT Press, 2002), 79–119.

12 James Der Derian, "International Theory, Balkanisation and the New World Order," *Millennium: Journal of International Studies* 20:3 (1991): 485–506.

13 Der Derian, "International Theory, Balkanisation and the New World Order."

14 Andrew Hammond, ed., *The Balkans and the West: Constructing the European Other, 1945–2003* (Farnham: Ashgate, 2004).

15 The term balkanization tends to be written with a capital 'B,' however, to continue with this prescription is to accept that there is a given universal truth about this discourse,

including the knowledge about the Balkans. Following Foucault's thinking, one needs to dispel the belief in a single Truth (with a capital 'T'). There is not a single understanding of Balkanization; instead, there are multiple conceptualizations and interpretations of balkanization.

16 Andrew Wear, *Knowledge and Practice in English Medicine, 1550–1680* (Cambridge: Cambridge University Press, 2000); Amy Swiffen, *Law, Ethics and the Biopolitical* (New York: Routledge, 2010); Michael Dear, "Los Angeles and the Chicago School: Invitation to a Debate," *City & Community* 1:1 (2002): 5–32; Alberto Alesina and Eliana La Ferrara, "Participation in Heterogeneous Communities," *Quarterly Journal of Economics* 115:3 (2000): 847–904; Douglas S. Massey and Nancy N. Denton, *America Apartheid* (Cambridge, MA: Harvard University Press, 1993); Peter T. Leeson, "Balkanization and Assimilation: Examining the Effects of State-Created Homogeneity," *Review of Social Economy* LXV:2 (2007): 141–164; Alejandro Portes and Robert L. Back, "Immigrant Earnings: Cuban and Mexican Immigrants in the United States," *International Migration Review* 14 (Fall 1980): 315–341; Alejandro Portes and Alex Stepick, "Unwelcome Immigrants: The Labor Market Experiences of 1980 (Mariel) Cuban and Haitian Refugees in South Florida," *American Sociological Review* 50 (August 1985): 493–514; Loic Wacquiant, *Urban Outcasts* (Cambridge: Polity, 2007); Howard E. Aldrich and Roger Waldinger, "Ethnicity and Entrepreneurship," *Annual Review of Sociology* 16 (August 1990): 111–135; John R. Logan, Richard D. Alba and Thomas L McNulty, "Ethnic Economies in Metropolitan Regions: Miami and Beyond," *Social Forces* 72:3 (March 1994): 691–724; Portes and Back, "Immigrant Earnings"; D. G. Shane, "Balkanization and the Postmodern City," in *Mortal City*, ed. Peter Land (New York: Princeton Architectural Press); Saskia Sassen, "Europe's Migrations: The Numbers and the Passions Are Not New," *Third Text* 20:6 (November 2006): 635–645; Aristide R. Zolberg and Long Litt Woon, "Why Islam Is like Spanish: Cultural Incorporation in Europe and the United States," *Politics and Society* 27:1 (March 1999): 5–38; William Kurt Barth, *On Cultural Rights: The Equality of Nations and the Minority Legal Tradition* (Boston: Martinus Nijhoff, 2008); D. Parthasarathy, "Decentralization, Pluralization, Balkanization? Challenges for Disaster Mitigation and Governance in Mumbai," *Habitat International* 52 (2016): 26–34; James Goudkamp, *Tort Law Defences* (Oxford: Hart Publishing, 2013); Jonathan H. Adler, "Climate Balkanization: Dormant Commerce and the Limits of State Energy Policy," *LSU Journal of Energy Law and Resources* 3:1 (Fall 2014): 153–194; Jonah Hill, "Internet Fragmentation: Highlighting the Major Technical, Governance and Diplomatic Challenges for U.S. Policy Makers," Report, Belfer Center for Science and International Affairs (BCSIA) – Harvard Kennedy School (HKS), 2012; John D. Woodward, Jr., "The Law and the Use of Biometrics," in *Handbook of Biometrics*, ed. Anil K. Jain et al. (New York: Springer, 2007), 357–379.

17 Some of the notable books that have established Balkanism as a critical study are Maria Todorova's *Imagining the Balkans*; Vesna Goldsworthy's *Inventing Ruritania: The Imperialism of the Imagination* (London: Hurst, 2013); Dušan I. Bjelić and Obrad Savić's *Balkan as Metaphor*; and Ludmilla Kostova's *Tales of the Periphery: The Balkans in Nineteenth Century British Writing* (Veliko Turnovo: St. Cyril and St. Methodius University Press, 1997).

18 In 1918, the country was known as the Kingdom of Serbs, Croats and Slovenes, to be renamed Kingdom of Yugoslavia in 1929. In 1945, the country was affirmed as Federal People's Republic of Yugoslavia but only a year later – in 1946 – constituted as the Socialist Federal Republic of Yugoslavia.

19 Macedonia also underwent some tensions in 2001, between the Albanian National Liberation Army and the Republic of Macedonia's forces (about 25 per cent of the population in Macedonia is Albanian Muslim, and 75 per cent Macedonian Orthodox). The insurgency lasted for nine months, and the casualties remained relatively low on both sides.

20 The revolutions of 1989 are also associated with the end of the 20th Century.
21 Some of these texts include: Morton Abramowitz, "Preface, The Other Balkan Wars: A 1913. Carnegie Endowment Inquiry," in *Retrospect with a New Introduction and Reflections of the Present Conflict*, ed. George Kennan (Washington, DC: Carnegie Endowment for International Peace, 1993); Charles Gati, "From Sarajevo to Sarajevo," *Foreign Affairs* 71:4 (Autumn 1992); Samuel Huntington, "The Clash of Civilizations?" *Foreign Affairs* 72:3 (Summer 1993); Robert Kaplan, *Balkan Ghosts: A Journey Through History* (New York: Vintage Press, 1993); Noel Malcolm, *Kosovo: A Short History* (New York: New York University Press, 1998).
22 Christopher Bennett, *Bosnia's Paralysed Peace* (London: Hurst, 2016).
23 It must be noted that during the Socialist period of Yugoslavia, despite the noted heterogeneity, religion was disqualified as being of national value and/or individual choice; its practice was largely denied and associated with extremist leanings. This was particularly the case up until 1960s–70s, when it became tolerated, as seen with the possibility to mark and participate in religious holidays. Also, the inability to differentiate from the socialist ideology meant that in the aftermath of WWII, individual right to land was abolished; the land was confiscated and nationalized.
24 Todorova, *Imagining the Balkans*, 3.
25 Der Derian, "International Theory, Balkanisation and the New World Order."
26 Walter Benjamin, "Theses on the Philosophy of History," in *Illuminations*, ed. Harry Zohn (New York: Schocken, 1977), 260.
27 Clinton's speech on television from the Oval Office on the evening of 24 March 1999.
28 Cited in Noam Chomsky, *The New Military Humanism* (London: Pluto Press, 1999), 5.
29 Noam Chomsky, *Failed States* (New York: Metropolitan Books, 2006), 101.
30 From an architectural and urban perspectives, connections between balkanization and the former Yugoslav context have been done by Kai Vöckler and Srdjan Jovanović Weiss. Refer to: Kai Vöckler, "'Balkanology' and the Future of the European City," in *Balkanology – SAM No. 6*, ed. Francesca Ferguson and Kai Vockler (Basel: Christoph Merian Verlag, 2008) and Srdjan Jovanović Weiss, "NATO as Architectural Critic," *Cabinet* 1 (2000/01); *Almost Architecture* (Stuttgart: Merz+Solitude, 2006); "School of Missing Studies," in *Did Someone Say Participate? An Atlas of Spatial Practice*, ed. Markus Miessen and Shumon Basar, 241–257 (Cambridge, MA: MIT Press, 2006); "Milosevic as Architect," in *Networked Cultures: Parallel Architectures and the Politics of Space*, ed. Helge Mooshammer and Peter Mörtenböck (Rotterdam: NAi Publishers, 2008); "National, UN-National," *Nationalities Papers: The Journal of Nationalism and Ethnicity* 41:1 (2013): 90–108.
31 Paul Hirst, *Space and Power* (Cambridge: Polity Press, 2005).
32 Weiss, *Almost Architecture*
33 Weiss, "School of Missing Studies," 241–257.
34 Weiss, "School of Missing Studies," 252–253.
35 Weiss, "School of Missing Studies," 256.
36 Gilles Deleuze, "Postscript on Control Societies," in *Negotiations* (New York: Columbia University Press, 1995), 181.
37 Stephen Graham, *Cities Under Siege: The New Military Urbanism* (London: Verso, 2010), xxix.
38 Chomsky, *The New Military Humanism*.
39 Referenced in James Der Derian's "Global Swarming, Virtual Security and Bosnia," *The Washington Quarterly* 19:3 (1996): 45–46.
40 Jagdish Bhagwati, *In Defense of Globalization* (Oxford: Oxford University Press, 2007).
41 Manuel Castells, *The Rise of the Network Society* (Oxford: Wiley-Blackwell, 2000).
42 Zygmunt Bauman, *Liquid Fear* (Cambridge: Polity Press, 2006); Zygmunt Bauman, *Liquid Times* (Cambridge: Polity Press, 2007).

43 Paul Virilio, *Speed and Politics: An Essay on Dromology* (Los Angeles: Semiotext(e), 2006), 7.
44 Virilio, *Speed and Politics*, 1.
45 Virilio, *Speed and Politics*, 21–22.
46 The Rambouillet Agreement was drafted by NATO in what was represented as a peace agreement between the Federal Republic of Yugoslavia and the Albanian majority population in Kosovo. Yugoslavia's refusal to sign the agreement – on the basis that it contained provisions for the autonomy of Kosovo as well as an attempt to control flows of communication and movement – became a justification for NATO's air strikes.
47 Susan Woodward, *Balkan Tragedy: Chaos and Dissolution After the Cold War* (Washington, DC: Brookings Institution, 1995), 29–46.
48 Michel Foucault, *The Archaeology of Knowledge and the Discourse on Language* (New York: Vintage, 1982), 23–24.

2 Balkanism and balkanization

Fragmentation, grouping and excess

The Balkans

Balkanism and balkanization

Historically, the term Balkanism has predominantly been associated with violence and sociopolitical instability. The concept is used to identify the Balkans and is laden with an imaginary of ethnic hatreds. Dušan Bjelić's position on the imaginary violence is that the works of prolific writers on Balkanism[1] tend to approach the discourse and the zone of the Balkans as a body of knowledge that straddles both the study of colonial representation (as distinct from Orientalism) and the genealogical question of how certain 'truths' arose about the Balkans.[2] Within this question are entangled aspects of place, language and history; indeed, even the political relation between image and identity in terms of geography and culture. Questioning the source and conditions of knowledge to do with the representations of the Balkans and Balkanism helps to destabilize the very construct of the Balkans as a barbaric zone. One contribution to this destabilization is to recognize that, prior to the 18th Century, the word *Balkan* in the English language was associated solely with a mountain range, a peninsula and a geological frontier between Europe and Africa.[3]

The first use of the word *Balkan* in Europe was in the 15th Century by the Italian humanist and diplomat Filippo Buonaccorsi Callimaco to note the Turkish name given to a mountain. However, it was not until the mid- to late 18th Century that *Balkan* became a common name for geographers and travellers; it was introduced into English language by Frederick Calvert Baltimore.[4] Thus, it had taken nearly four centuries for the Turkish word *Balkan* to gain currency in the West; the Ottomans had been using it since the start of their occupation of Europe in the 1300s. It is possible that at that time the term was used to describe the geography of the area, since in Turkish, the word *bal* means 'mud,' and *kan* is the diminutive.

It is only between the French Revolution in the late 18th Century and the collapse of the Ottoman Empire in the late 19th and early 20th Centuries that the modern Balkans emerged on the political map as a region. The bringing together of areas under the Ottoman Empire in this way had been proposed by the German geographer August Zeune in the early 19th Century. The imaginary of the

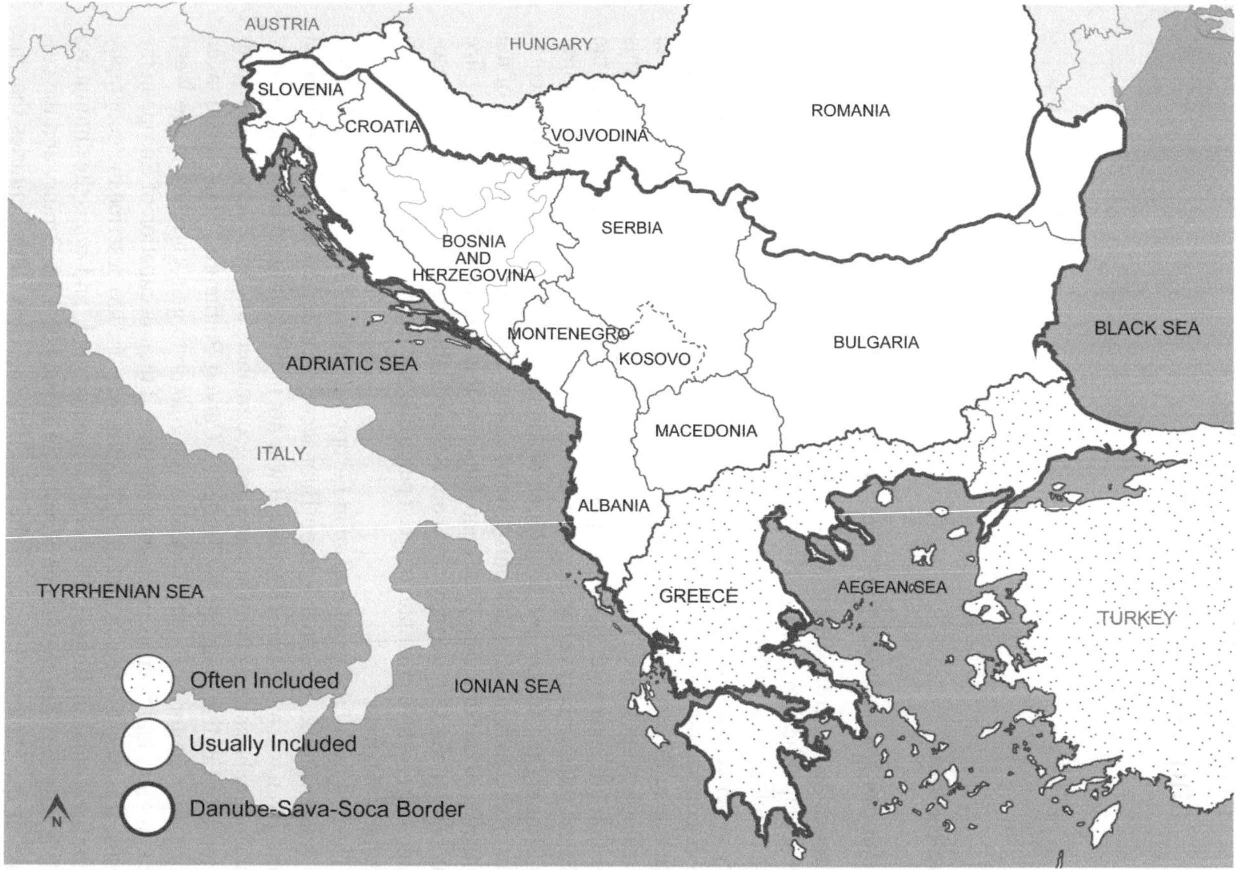

Figure 2.1 The different configurations of the Balkans – as an often included and usually included political entity, to its actual geographical Danube-Sava-Soca boundaries.

Balkans as a passionately violent zone was attached only after the Ottoman Empire withdrew; the withdrawal process coincided with the 1878 Treaty of Berlin. The territories that were until recently under the Ottomans were balkanized as the ethnically pure nation-state model was superimposed there. While the ideal of a homogeneous nation state developed gradually in Western Europe, it was imposed politically relatively abruptly in the Balkans. This is notable since the region has been acknowledged historically as having the greatest ethnic heterogeneity of cultures, languages and ethnicities in Europe.[5]

That diversity largely derived from its geographical position in that it has historically been at crossroads for trade, communications and military campaigns starting with ancient Greece and Rome and continuing across Byzantium, Ottoman Turkey and Roman Catholic Europe. The ethnically purist Western initiative imparted onto heterogeneous nations led, unsurprisingly, to violent civil unrest;[6] events which earned for the Balkans a reputation as a region of endemic violence inhabited by semi-civilized and only semi-European barbarians.[7] The recent balkanization of Yugoslavia in the 1990s further cemented the image of the Balkans as a rogue region and a chaotic frontier, one whose racial and ethnic disagreements have always been deep-seated and thus predestined for eternal conflict, since 'civilized ways' are not possible there.[8] However, balkanization as a method of fragmentation has a significantly longer history.

The term balkanization was first mentioned in the 1918 *New York Times* article "Rathenau, Head of Great Industry, Predicts the 'Balkanization of Europe.'"[9] The interviewed Dr. Rathenau described the political, industrial and economic calamities found in Germany post WWI potentially having devastating effects on Western civilization should Germany not take wake up and take charge; Germany, indeed, was potentially about to lose further territory and power.[10] Another series of journalistic reflections were made in 1921, the early 20th Century fragmentation of European empires and the trail of dissent, ravages of WWI and the peace treaties were described as "Balkanized Europe."[11] The particular focus being on the newly formed yet also backward, financially weak, distrustful and often too passionate parts of Eastern Europe and the Balkans, which were seen to require a long political education in order to understand the meaning of Western civilization and democracy.[12] The broad effects of "Balkanized Europe" were those of introversion, lack of dialogue and willingness to collaborate and as such hindering the possibility for economic growth. For the journalist Mowrer, the plausibility for Europe to unite and to follow in the footsteps of the US was not a possibility in the foreseeable future; Europe was seen as too ignorant, careless and immature.[13] Eric Hobsbawn described balkanization as formation of mini-states underpinned by nationalism and associated balkanization with the fragmentation of the Balkans after WWI.[14] However, as Todorova correctly points out, "all Balkan states, except Albania, had existed from several decades to a century before World War I."[15]

Balkanization as a geopolitical *act*, *method* and a *process* was spurred by the withdrawal of the Ottoman Empire from Europe. The move was instated for purposes of modernization and attempts to achieve ethnic homogeneity within each territory. The need to extend the values of the Enlightenment complemented the desire to enlarge the footprint of Europe, though only if the land incorporated retained a subordinate imaginary. This was seen in the latter part of the 19th Century when all the areas still under the Ottoman Empire were referred to as European Turkey. Balkanization is, thus, not only implicated in borders but also in nation shifting, where newly created neighbouring territories may not only be hostile towards each other but where maintenance of peace and order is ultimately dependent on the West. This was seen with the 1878 Congress of Berlin and has persisted to this day. In the heat of NATO's 1999 targeting, over 120 years after the Congress of Berlin, Robert Kaplan suggested that "[o]nly western imperialism – though few will like calling it that – can now unite the European continent and save the Balkans from chaos."[16] It seems both that the Balkans are 'at a lower evolutionary scale' and that to evolve is dependent on external guidance lest violence should ensue.[17]

Balkanization is thus implicated in the Balkanist imaginary where Europe is seen as civilized, reasonable and tolerant while the Balkans are portrayed as a place of wilderness, irrationality and unending conflict. Seemingly, balkanization is not just a question of Todorova's thinking about Balkanism in terms of identity reversal of the Balkans[18] – from backward and primitive to progressive and developed – but also for two other agendas. The first is that this one part of the world has historically never been developed and civil, and that it needs to be 'trained' to become so. Second, it is held that the need for this training is limitless since history purportedly shows that violence persistently erupts in the region despite – oddly – the historic intervention of 'tried and tested' democratic policies that have made the West an enduring beacon of civility and peace.

Even after becoming a part of the European polity through its separation from the disintegrating Ottoman Empire, the violence and barbarity associated with the Balkans have persisted; this resistance to the civil values found in Europe and the West has been put down to its "semi-European" nature.[19] These representations are a political construct imported from outside, from Europe and the West. According to Vesna Goldsworthy, the Balkan Peninsula comes to represent some kind of "anti-Europe"; it is always "not yet European" or "that which Europe has been long ago."[20] Todorova frames this in-between connotation within a perspective that holds that the Balkans are a part of Europe though on its periphery.[21] For Aleš Debeljak, the geographical and territorial reading of the Balkans is embedded in mythological symbolism and associations that yoke together a rational, progressive and technical Europe and a fanatical, underdeveloped and primitive East.[22] The implication is that the Balkans are a space with a transitory character, where time operates in ways that are outside European measures as well as outside the preferred European, or Anglo-American, norms.[23]

The aesthetic of the unruly imaginary of the Balkans has pervaded other fields of arts and culture too, with film and literature taking centre stage in driving this perception and maintaining it, particularly its association of the Balkans with semi-civilization, violence and monstrosity. Goldsworthy has charted the Balkanist prejudices found in works of English literature such as Bram Stoker's

Dracula (1897), Anthony Hope's *The Prisoner of Zenda* (1894) and Olivia Manning's *Balkan Trilogy* (1960), amongst others.[24] One of the characters in Agatha Christie's *The Secret of Chimneys* (1925), the peasant Boris Anchoukoff, is said to be the epitome of a man from the fictitious small Balkan country Herzoslovakia, having "high Slavonic cheekbones, and dreamy fanatic eyes" and being "a human bloodhound from a race of brigands."[25] Stoker's *Dracula* is also noteworthy as it traverses both literature and film (dir. Bela Lugosi, 1931). Within the novel, the origins of Dracula are found in the exotic Balkan land of fear and blood. For Bjelić, the imaginary of *Dracula* offers insights into the genealogy of violence in Europe more broadly in that the Balkans as a zone of enduring violence is coeval with the association of Europe with enduring peace.[26] Such proportionality is no longer just a question of Goldsworthy's writing on Balkanism in terms of "imperialism of the imagination,"[27] but is now a matter of imperialism of the aesthetic that raises questions about the intellectual landscape of those who succumb to the constructed aesthetic. Whilst some of the characterizations attributed to Balkanism may sound Orientalist, Balkanism and Orientalism display important differences.

Scholars of the Balkans differ on the exact relation of Balkanism to Orientalism. According to Todorova, while the discourse of Orientalism is about asserted oppositions, the discourse of Balkanism is about asserted indistinctness.[28] In terms of subordination, for Elissa Helms, Orientalism is directly connected to Western colonization, while Balkanism arose in less direct ways; Western impositions of domination and expectations of subservience were more diffuse.[29] For Močnik, Balkanism has inherited Orientalist legacies, though configured twice: as the myth of European progress, and as the myth that backwardness is inherent in non-European contexts.[30] Other authors on Balkanism, such as Robert Hayden and Milica Bakic-Hayden, position Balkanism as a variation on Orientalism in view of the centuries-long Ottoman rule of portions of the Balkans; the belief held by Hayden and Bakic-Hayden is that the region is irredeemably tied to the Orient.[31] The variant is described as a process of "nesting Orientalism" within the zone of the Balkans itself, whereby each region in the zone would identify areas south or east of it as having traces of Orientalism or primitivism.[32] Bakic-Hayden and Hayden's thinking complements the symbolic imaginary found in the thinking of Said, for whom the Orient is a project of essentialist stereotypes rather than an actual place.[33] According to Močnik, Balkanism should be considered in terms not only of the zone's Ottoman history but also of the relation of Balkanism to globalization, from the perspective that post-1990s balkanization is contingent on inscribing this identity with the glossy image of globalization.[34] Indirectly, balkanization of the Balkans is further tied to eliminating the Balkans' association with communism and socialism.

The differing opinions are essentially driven by geopolitics. Definitions of the Balkans, straddling the Orient and the Occident, are simultaneously "semi-Oriental" and "semi-European," suggesting both spatial and representational ambiguity as well as transgressing and disturbing the conventional identifications of space and borders. Unlike the critical discourse of Orientalism, which arose in the Western world, the study of Balkanism is predominantly located within the Balkans. The imaginary of Balkanism refers to a concrete place with clear geographical outlines and historical certainty, by contrast with the figurative projections about the Orient.[35] Its defined boundaries, though, were pointing to areas that were under the Ottoman Empire and as such still symbolically alien to the rest of Europe. What the projections of Balkanism and Orientalism do share is their binaries of opposition: Western order, justice and civility against the Balkanist or Orientalist disorder, corruption and barbarism. Balkanism and Orientalism are always set up against an ideal and cast within a position of lack; what it means to be not part of Europe and the West in terms of values and norms.

Balkanization beyond the Balkan borders

Despite balkanization being used to parcelize the Balkans for purposes of eliminating the imaginary of this region as a zone of complexity and chaos – including its sense of difference from Western Europe and Anglo-America from the perspective that the Balkans as a term still carries derogatory connotations of volatility in the lifestyle and behaviour of its inhabitants – the application of balkanization has branched out.

That balkanization has been extracted from the geographical context of the Balkans, and adopted elsewhere, with at times less volatile uses, is seen from the term's emergence into fields that include urban planning, urban sociology, law, digital communications media, medicine and biology, with varying currency and degrees of power-oriented connotation. In biology, the term has been used to separate systems from a common territory, thus establishing clear hierarchies. For example, there is a separation of molecular and evolutionary biology, or biochemistry; such a specialization-oriented method is reductionist as it does not treat the body as an integrated system but one of divided parts. This separation is also evident in the 16th and 17th Century parcelization of the field of medicine by distinguishing medicine from its supporting areas such as nursing, the subdivision weakening the field.[36] The negative implications within medicine do not end there, since the greatest perceived risk is in the way in which morality and ethics are cognized in specialist fields such as neurology and cardiology. Subdivision parcelizes the very consideration of ethics; action is allowed upon certain isolated concerns without considering the human body as an integrated and complex organism.[37]

The application of the term balkanization, however, was predominantly used to describe situations in the 20th Century. In the 1940s, the term was used in passing and in respect to projected economic consequences to draw attention to the negative economic effects likely to flow from blockage of interstate commerce and trade. In the 1960s, Austrian and German institutions were characterized as corrupt, segregated and mismanaged.[38] The relation between institutions and segregation is particularly prominent in colonial territories. The period after WWII and into the 1960s saw the use of balkanization to predominantly describe the decolonization of Africa. In the 1960s, the English and the French described the newly independent states in derogatory terms from the perspectives that the states were facing a bleak and frustrating future, and one which was bound to turn into a version seen in the hostile Balkans.[39] The alternative to this view is that balkanization in Africa was implemented for purposes of economic interest and Western power.[40] More recently, balkanization has been used to draw relations to the processes of bureaucracy. For example, the British model of bureaucracy has left its mark in once-colonial territories, predominantly when it comes to government and legal systems. In Mumbai, measures to protect the environment and the ecosystem in the event of a disaster such as a flood are significantly less effective due to the balkanized structure of city administration originally bestowed by the British; there is a breakdown of dialogue between different bureaucratic domains, which then immediately hinders the efficiency of any disaster mitigation.[41]

The polarizing effects of balkanization are most deleterious when the outcome has the capacity to directly affect people. At that point, it no longer just describes a context but operates as a means to achieve fragmentation. When it comes to urbanism, balkanization has been invoked in situations of security, segregation and conflict. In Octavia Butler's futuristic 1993 novel *Parable of the Sower*, balkanization is used to describe a divided population some of whom live in gated communities while others do not, a situation that leads to wars in and around Los Angeles.[42] The dystopian novel projects a world of extreme privatization, disparity of wealth and protectionism.[43] This fictionalized narrative complements the real-life urban concerns aired in the writings of Michael Dear. Dear describes a proto-postmodern condition driven by global restructuring and market economics, the outcome of which is a series of balkanized enclaves distinguished from each other socially and culturally, yet also politically and economically polarized, as evident in the existence of fortified and peripheral cities. These ideas are also addressed in Mike Davis' writings, particularly in terms of lack of dialogue between different socioeconomic and ethnic zones exacerbated by the presence of enclaves.[44] This can be contrasted with Douglas Massey's and Nancy Denton's readings of US spatial politics, namely that enclaves are created to achieve residential racial segregation,[45] which, in turn – according to Leeson – contribute towards establishing a business network and "enclave economies."[46] These enclave economies also contribute towards creation of personal networks.[47] However, the fact that these enclave economies are inextricably linked to "ethnic enclaves"[48] means that assimilation and the possibility for coexistence in the US not only is limited but also facilitates conditions that over time can turn into physical and spatial incursions. When balkanization is used as a method for purposes of coexistence, it facilitates opportunities for heterogeneity.

That the fictional agenda of Butler's text became a measure of reality was seen in US President Bill Clinton's likening of the 1992 riots in Los Angeles, California, when a white policeman beat up the black citizen Rodney King, to the ethnic cleansing spurred by racial and ethnic differences encountered in Bosnia and Herzegovina, particularly the violence between the Serbs and Muslims.[49] Drawing parallels between the riots and the Balkans was a way to absolve American institutional and urban practices of responsibility for the polarization and privatization of wealth. The Balkans were invoked to imply that this type of violence is fundamentally inconsistent with the origins and identity of America. What went unacknowledged in Clinton's comparison is that the riots arose out of the disparity of wealth and access to resources as well as the effects of gated communities on American urban planning.

Balkanization is used as an urban strategy for purposes of creating enclaves, often on the periphery and in order to subvert or maintain order; the outcome of which for D.G. Shane is equivalent to the qualities of a heterotopia.[50] They can be deployed to effect the devastation, invasion or fracturing of communication.[51] These balkanized enclaves and their interstitial spaces are similar to the ravages of Sarajevo; they can easily attain the violence seen in the Balkans, according to the perception of the former Yugoslav model as the example par excellence of paramilitary gangs and petty dictatorships.[52] Balkanization, from an urban planning perspective, is tied to not only questions of culture and ethnicity but also geography and policing of mobility, which Saskia Sassen frames in terms of migration and geopolitics.[53] Questions of migration may also be seen as a carefully engineered programme where, along with the restriction of welfare rights, there is also a possibility to control racial diffusion or the "browning of America."[54] It is a discriminatory regime that is evident not only in the US but also in Western Europe and Australia.

The most recent UK example was the campaign in support of Brexit. The dominant media rhetoric was built not around critical analysis of EU policies but a fear that the influx of migrants was having a negative impact on the UK economy and the British citizenry. Here, the threat was tied to underlying questions of money even though the surface agenda was that the UK should retain an authentic identity for which land, language and people need to remain in alignment; rhetoric reminiscent of Martin Heidegger's claim that a people and their land are a unity via language.[55] In other words, there is a connection between a people, their land and their language; everyone who does not belong to that soil, whose being does not arise from a distinct place or cannot trace their history in terms of that language and land is automatically defined as Other and outsider. It is the argument of racial hygiene; one that is very similar to the rhetoric found in Nazi ideologies constructed around racial superiority and the perceived threat posed by the Jews. It has been seen, in diluted form, even within the UK's borders. UK planning laws remain discriminatory towards the Romani; they are directed towards the expulsion of the Romani from their caravan settlements, as well as their segregation.[56]

Where balkanization could be said to have the greatest impact in legal processes is during trials where overlapping jurisdictions are involved, making it harder to receive justice, an outcome that is particularly visible in colonial territories such as Australia and the US.[57] For example, when addressing aspects to do with compensation in terms of civil rights such as property rights or rights to one's health, safety and a clean environment, the dispensing of justice becomes harder since each jurisdiction operates under its own set of laws. Thus, justice becomes more elastic and easier to manipulate precisely because the administration of law becomes more bureaucratic. In terms of the environment, balkanization is also implemented in matters to do with energy regulation, ranging from control of fees to developing more environmentally friendly energy sources.[58] In other words, balkanization is used as a legal policy to "enact environmental regulations and otherwise control the environmental effects of energy use and production within their borders."[59]

The legal, economic and political insecurities afflicting Western Europe and Anglo-America are also affecting the way in which national identity is comprehended in that it is tied to the colonial mentality of ownership and territorial expansion. In Australia, such colonial motivations have been in operation since its post-1788 Anglo-Saxon colonial formation, tied as it was to the elimination of Other peoples, values, languages and customs through the genocide directed towards the indigenous Aborigines. Here, balkanization was implemented by fragmenting the conceptions of land and territory that accorded with indigenous beliefs and superimposing borders that discriminated against indigenous values and customs.

Inceptionary 1788 colonial practices are seen to this day in the detainment of refugees on an archipelago of off-shore locations in the Pacific. Unlike the post-1788 colonial practices, those of the post-1990s are entrenched not only in the automatic detainment of all undocumented migrants arriving by boat, but also in the legitimization of market colonialism. The economic agenda is present because vast amounts of money have been invested in privatizing and sustaining the industry of border control as well as due to the rhetoric that migrants would supposedly erode the Australian economy, hence the need to detain all undocumented refugees in off-shore camps. One of these is located on Christmas Island, a nature reserve also mined for its rich deposits of phosphate, that has since late 20th Century been used for purposes of imprisoning 'illegal' immigrants. Purchased by the Australian government from the British in the late 1950s, the terror exercised on this island has largely remained hidden from the mainstream media.

That these practices have been appropriated and exported by European governments in the name of security – that is, market economics – is one of the more twisted sequelae of addressing matters to do with 'illegal' migration. The benefits of territorial expansion, if done through policy alone, are that peripheral zones can now be used for purposes of containing the migrants and refugees. This is seen in the agreements struck by European governments, inclusive of the UK, with countries such as Libya and Turkey for purposes of detaining asylum seekers.[60] These peripheral zones are a more isolated and regimented version of the 18th Century peripheral institutions such as prisons where treatment of inmates and disciplining methods were removed from visibility. The disciplinary measures are entangled in doctrines of risk, security and legal exceptionalism.[61] With United Nations High Commissioner for Refugees (UNHCR) 2018 records indicating that 68.5 million people have been forcibly displaced,[62] and as the design of camps often comes to mimic that of prisons, to be displaced and in search of a new home is equivalent to having breached the law. Balkanization is present upon the homogeneous stamping of the displaced with an imprisoned identity of never being able to belong; their existence dependent upon exclusion and separation.

Understanding territory is particularly peculiar in a field hailed, since its inception, for its openness and accessibility to all: the internet. The balkanization of digital communications media through the creation of distinctive enclaves is driven by aspirations to ensure data sovereignty. One key example of the need to establish sovereignty of data was seen in the 2013 case where Edward Snowden was able to reveal that US classified data had been shared with various European governments and companies without authorization.[63] Balkanization of data and privacy in the digital age occurs in at least two ways. The first is to do with the growing range of biometric identification tools (from fingerprint scans to voice recognition), which is known as biometric balkanization.[64] Here, the use of the term balkanization tends to be seen in a positive light, not only because the increase in biometric tools allows a more optimal match for a specific need but also because compartmentalization of these tools facilitates the alignment of individual desires and organizational interests,[65] the effects of which – however – may lead to not only conflation between the individual and the organization but also open manipulation and control of the former by the latter. The second aspect of data and privacy is to do with the balkanization of the internet for purposes of security and censorship, by – for example – keeping data within EU boundaries protected from other territorial zones such as the US internet infrastructure networks,[66] which would have notable implications for trade and financial flows.

Balkanization and Yugoslavia

The broad framing of the Balkans and Balkanism, including the use of the term balkanization globally and in diverse fields, was necessary in order to appreciate the significance of Yugoslavia's implementation of balkanization and this region's specificity within the Balkans. The constructed violent imaginary of the Balkans tends to eclipse the ethnic heterogeneity and national constituency found within the formation of Yugoslavia in the 20th Century, which continued the lineage of diversity historically found in the Balkans.

The path to the 1918 formation of a heterogeneous Yugoslavia through unification of the South Slav peoples into one nation state was paved with violent nationalist outbursts. For Yugoslavia, a prerequisite for national unification was the removal of Austro-Hungarian rule. One significant and widely noted example of such violence occurred in 1914, when a Bosnian Serb nationalist Gavrilo Princip (in)famously shot and killed the Austrian Archduke Franz Ferdinand and his wife Sophie, Duchess of Hohenberg, in Sarajevo. Princip's act played a role in the imminent outbreak of WWI, but it was aimed as a blow against the Austro-Hungarian rulers, who met any opposition and especially expressions of anarchy with murder and deportation to concentration camps.[67] An act of anarchy intended to unite the South Slavs also confirmed the reputation of the Balkans as a despotic region, one that so soon after freeing itself from Ottoman rule was already pulling away from supposed European civility; it was turning back towards its 'originary' powder-keg barbarism.

Figure 2.2 Memorial in Sarajevo to mark the 1914 assassination of the Archduke Franz Ferdinand of Austria and his wife Sophie, Duchess of Hohenberg.[68]

The despotism of the Balkans belonged to a different type of European culture. This Other European is the black sheep of the family, a threatening anomaly to be kept apart from, yet a part of, the ordered and unified European whole; the abnormal region forever in need of taming. It may be said, however, that the 1914 assassinations were not just a blow for national self-determination, self-rule and unification of culturally close groups but, more important, were an expression of resistance to the way Western interests privileged a certain idea of homogeneity in their prescriptions for the future of a region of long-standing heterogeneity.

The desire to unify the South Slavs (apart from Bulgarians) and the culturally close groups that lived in Slovenia, Croatia, Serbia, Bosnia and Herzegovina, Montenegro and Macedonia was given concrete expression in 1918 with the formation of the Kingdom of Serbs, Croats and Slovenes (renamed the Kingdom of Yugoslavia in 1929, even though the name Yugoslavia was used colloquially since 1918).[69] This first Yugoslavia also incorporated non-Slavic minorities such as Albanians, who were largely found in the province of Kosovo and, to a smaller extent, in south Serbia, Montenegro and Macedonia. The peoples who lived in the region were culturally aligned through their common Slavic ancestors who, over time, separated into three different tribes who, in the 7th Century, settled in different parts of the southern Balkans. With this in mind, the reverse balkanization of 1918 may be seen as a reunification of the three Slavic tribes.

However, with so much history between the initial settlement, tribal separation and reunification in the 20th Century, the parliamentary system established in the kingdom needed to reflect that these were separate nations with some 13 centuries of separate history despite their common cultural and linguistic roots. Croatia and Slovenia had been largely under Austro-Hungarian influence, but for Bosnia and Serbia, influence had come from both the Austro-Hungarian and Ottoman empires. Over the period, it is not surprising that other rulers too passed through or took territorial ownership of the noted areas. With the reorganization of the Balkans in general, and Yugoslavia in particular, it was also necessary to select a common tongue. In 1850, during the Vienna Agreement between Serb and Croat scholars, the Stokavian dialect was selected from three options (Stokavian, Cakavian and Kajkavian) and endorsed as the foundation of the common Serbo-Croatian language. This language became the standard in Croatia, Serbia, Bosnia-Herzegovina and Montenegro, with slight variation, as well as the significant difference that the official alphabet to be used in Croatia and Bosnia-Herzegovina was Latin, and Cyrillic in Serbia and Montenegro.

The Kingdom was structurally premised on each state having an independent financial system and its own body of laws as well as a constitutional power of veto. Perhaps as a consequence of centuries of incorporation in different empires and under different rulers, the polity of the first Yugoslavia inherited numerous different legal, monetary and transport systems.[70] The process of reconfiguring these administratively privileged the Serbs, due both to Serbia having a longer history of independence and the number of districts with majority Serb population located in the rest of Yugoslavia. On the other hand, republics that had historically been within the Austro-Hungarian Empire (such as Slovenia and Croatia) were more developed in terms not only of transport networks, but also economically. Here, market farming was developed and economics used both for local investments and infrastructural development as well as trade.

Areas that were under the Ottoman rule, such as Serbia, practised predominantly subsistence farming; this had an impact not only on their economy, but also on urban development and everyday life. The architect Le Corbusier's impressions of Belgrade seemed to confirm a city of chaos and excess. During his travels in 1911, he pronounced Belgrade to be an undefined city and "a ridiculous capital, worse even: dishonest city, dirty and disorganized."[71] These impressions formed part of the journey he undertook with his friend August Klipstein that had Constantinople as its final destination. Prior to arrival, Le Corbusier imagined Belgrade as "the magic door to the East, swarming with colourful life, populated by flashing and bedizened horsemen wearing plumes and lacquered boots!"[72] In the event, his records largely indicate less than complimentary impressions; the people in Serbia are described as ruddy, their speech as lacking zest, the wine as cunning and food as being too heavy.[73] At the time, Belgrade was clearly a mish-mash of different styles, including influences of Ottoman architecture. Neither the city nor Serbia lived up at that time to the expectations of Le Corbusier's purist modernist palette.

The formation of the second Yugoslavia began during WWII, during the Second Anti-Fascist Council for the National Liberation of Yugoslavia convened in the Bosnian town of Jajce (Little Egg).[74] The overarching agenda of this second Yugoslavia was not only to retain the multiethnic palette but also to diversify it through the instatement of social, cultural and linguistic rights for its constituent nations (Serbians, Croatians, Slovenians and Macedonians); the Bosnian Muslims would later be recognized as the sixth constituent nation in 1961. Such rights were also, over time, given to Albanians, Hungarians, Slovaks, Vlachs and others, though these remained national minorities. Territorially, this was evident with the formation of two provinces within Serbia after WWII: Autonomous Province of Vojvodina, with a notable majority of the Hungarian and Slovak national minorities, and Autonomous Province of Kosovo and Metohija, with a notable majority of Albanians living there. The conferring of rights was a socialist version of democracy, one that was – theoretically at least – strengthened through self-management, the intent of which was, through decentralization and subdivision in any working and social sector, to allow those participating in that sector to reap the fruits of their labour. The ultimate goal was for the state to wither away by transferring state power to the society.[75]

The 1990s disintegration of Yugoslavia was at the same time the dissolution of this model and the reintroduction of the barbaric imaginary. 1990s balkanization was a five-stage process of small wars. The 1991 war in Slovenia reached a conclusion relatively peacefully and with little legacy of physical destruction or casualties. The following second and third stages were the Croatian War of Independence (1991–95), and the Bosnian War (1992–95); these were wars of physical erasure, uprootings of heterogeneity and solidarity. The following fourth stage was NATO's 1999 'humanitarian' intervention. The current, and fifth stage, is tied to what has taken place since each of the prior conflicts was terminated. This stage has been evident not least in the official renaming of the region as Western Balkans,[76] a term adopted in Brussels in 1998. By grouping together all remaining unruly and 'problem' countries, Western Balkans is no longer a geopolitical categorization but a political one. The EU membership is likely to remove the political implications and associations of the country being Balkanist.[77] For the EU, the Western Balkans comprise Albania and, except for Slovenia, all the republics of former Yugoslavia. The European Bank for Reconstruction and Development (EBRD) uses the term Western Balkans to refer to all countries recognized by the EU, except for Croatia. These unruly Western Balkans areas still carry the Balkanist slur of an imaginary "ghetto enclave" of violence.[78] This helps explain why entry into the EU for successor republics of the former Yugoslavia has generally been subjected to questions to do with liberal democratization and whether the extremity of 1990s violence is acknowledged as ethnic cleansing and genocide.

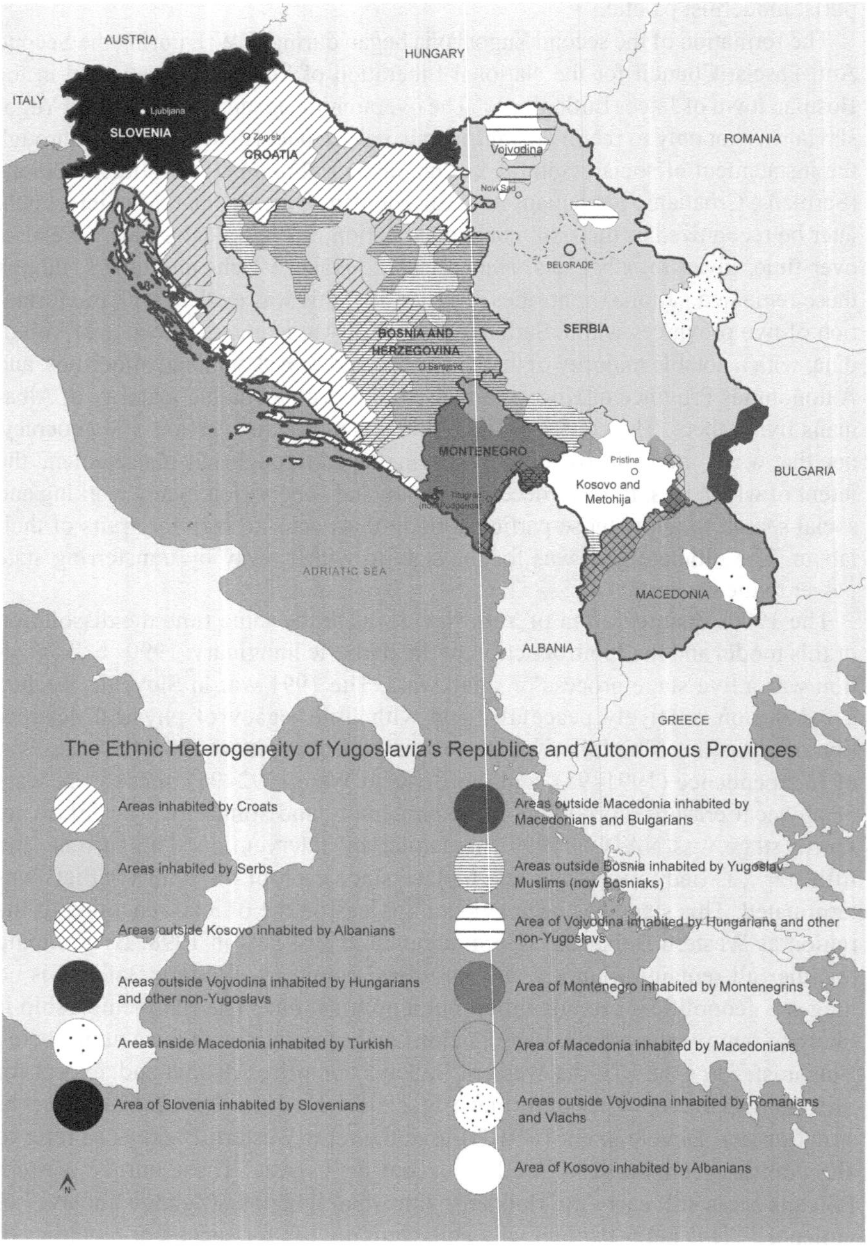

Figure 2.3 The ethnic diversity of the Socialist Federal Republic of Yugoslavia (SFRY).

Figure 2.4 The constitution of Western Balkans countries. Croatia is occasionally included, depending on whether it is defined by the EU, or the European Bank for Reconstruction and Development.

However, the pressing slur of violence was selectively mapped when granting membership to some former Yugoslav republics. Croatia's unclear Western Balkans identification may explain its recent EU membership despite the fact that it has since the 1990s become one of the most ethnically pure countries in Europe by exterminating and/or converting to Catholicism its once constituent nations of different faiths and ethnicities.[79] The preoccupation with ethnic purity is particularly ironic for this region considering its historical heterogeneity. Croatia's links to Nazi Germany and mass genocide during WWII also seem to have been conveniently forgotten by the EU.[80] This is not to deny the atrocities committed in Bosnia and Herzegovina by Bosnian Serbs (as widely reported in the news media), or those committed by Bosniaks – Bosnian Muslims (somewhat less reported). It is more to put into question the sanitation of particular acts of violence and the construct of tolerance practised by Europe. However, tolerance is oddly situated in the now European Croatia considering that its late 20th Century formation was contingent on the forcible expulsion of about 200,000 ethnic Serbs as part of its 1995 Operation Flash and Operation Storm, otherwise known as Homeland War. Similarly, the supposed progressiveness is also contingent on averting one's gaze from the trauma of concentration camps for Jews, Romani and Serbs during WWII when Croatia was a Nazi puppet State. The mission to cleanse this history from the Balkans selectively decides which countries are allowed to erase that violence in order to cross the European threshold of stability, peace and democratic dialogue.

Balkanism and balkanization as heterogeneity

Public space

Addressing Yugoslavia, Balkanism and balkanization genealogically, however, undermines the very potential of this region of frictional liminality as well as that of the country that took a counterapproach to balkanization by affirming and expanding its heterogeneity. What the genealogical turn, however, does offer is the opportunity to return to the conditions that brought the construct into a reality. From a different perspective, the discourse of Balkanism may be reconceived by deploying the thinking of Paul Patton, for whom revolutionary action is contingent on the very necessity to reconsider the text and its interpretation.[81] This reconsideration is contingent on amplifying Balkanism and Yugoslavia's ambiguous and heterogeneous potential, since both are signifiers of what Europe casts out as abnormal and full of excess. Once its alternative, anomalous and abnormal identification, no longer has negative associations, opportunities then arise to move history beyond the frozen frame of Balkanism as villainy.

Where the spatialization of Balkanist abnormality exposed and challenged national and international political inconsistencies to do with violence was evident during NATO's 1999 targeting, when Belgrade's public spaces hosted gatherings dubbed Songs Sustained Us (*Pesma Nas je Odrzala*), convened as a means not only to resist NATO bombings but also to reinhabit and reinterpret those spaces under attack. Many read the Songs Sustained Us events as yet more evidence of Balkan irrationality, including the former US Secretary of State Madeleine Albright, who responded, "I just don't understand anything, these people are totally insane."[82] The sites of the Songs Sustained Us gatherings included a significant portion of Belgrade's Old City centre, utilized not only for music performances, but also art exhibitions, film and sporting events. The gatherings were evocative of potential identified in Deleuze's thinking whereby pure events escape the present by moving in several different directions simultaneously.[83]

Notably, the Cultural Arts Society (*Kulturno Umetnicko Drustvo*) presented ensembles of Serbian traditional dance under the slogan "With dance and song we are defending Yugoslavia." Photographic exhibitions such as the conceptual photography of Dimitrije Cipcic's "Public Revolt" adorned Knez Mihailova Street. Artists brought out their canvases and painted in the streets. While some of the paintings had religious connotations, others were more graphic, with slogans such as "You target me with a bomb, and I target you with a paintbrush." The National Theatre, a building that fronts the Republic Square, staged numerous plays and operas. The theatre was considered a shelter for freedom of spirit, with performances taking place even at night while the bombs were falling on the city. Other events in Belgrade's public urban spaces included the 46th Festival of Short Film, the oldest film festival in Europe, which ran 14–19 April 1999. The Belgrade Marathon, which took place on 17 April, was organized before the actual bombings started, with confirmed participation by entrants from 40 countries. The event was not postponed, despite the withdrawal of international runners. The marathon was accompanied by the slogan "No bomb can break the soul and will of the capital city," and started appropriately in the then Boulevard of Revolution. The official poster for the race read "Stop the war, run the world."

All these events, which were a proclamation of the will to live, came to be labelled as acts of abnormality. Unlike D.G. Shane's inscription of heterotopias for purposes of establishing peripheral enclaves, the Songs Sustained Us events are more transgressive in that they reflect Foucault's notion of heterotopias; places where, as he states, "without boats, dreams dry up, espionage replaces adventure, and police [take the place of] the pirates."[84] The heterotopic spaces of Belgrade during the NATO air attacks maintained the dreams of the people and provided a way of transforming various national and international disciplining structures.

While NATO tried to extinguish Belgrade's economy, infrastructure and will, another minor economic and creative process began to take shape. Fashion parades were organized, with the black-and-white military target symbol finding its place on T-shirts and dresses and becoming a fashion statement. Street vendors and food stores became the "pirates" as they sold target buttons, posters and badges during the NATO bombings. The gatherings demonstrate that the body can still be integrated into historical processes whereby history is opened up through the manifestation of a will that extends beyond the governing set by national and international impositions.

These events were held despite the Pulitzer Prize–winning journalist for the *New York Times*, Thomas Friedman,[85] writing that "[i]t should be lights out in

Belgrade: every power grid, water pipe, bridge, road and war-related factory has to be targeted. [. . .] You want 1950? We can do 1950. You want 1389? We can do 1389 too."[86] Friedman's comments, according to Stephen Graham, urged not only that all that facilitates urban life be suspended and brought to a halt, but also that immobilizing movement and eliminating all resistance in the society is a question of reversing time by choosing the appropriate target and weapon.[87] Of the dates selected, 1950 is associated with the period of socialism in Belgrade (and Serbia), and 1389 is significant as it marked the defeat of Serbia by the Ottomans (as well as the loss of the territory of Kosovo).

Between Albright lamenting the apparent abnormality of those protesting and Friedman drawing attention to Balkanism as not just a permanent identification of this country but also the very means by which its prior history can be eliminated through time reversal, there is a construct that those who live in this region are somehow less human, thus implicitly validating the acts of 'humanitarian' violence directed towards the people living there as part of NATO's 1999 operation. Through the invocation of Balkanist abnormality, what the international community is doing is not an outright declaration of 'us' and 'them' but a reinvoking of the imaginary of the irredeemable barbarity present within the region and amongst the people. It is a process that automatically absolves Anglo-America and Western Europe of the commission of any act of violence, since violence occurs only in the Balkans. The technique is reminiscent of Carl Schmitt's writing on the political whereby politics is founded upon the mere identification of the enemy, which not only justifies but also validates the negative speak.[88]

The role of the human body during the 1999 protests suggests that bio power, management of the population, was not entirely possible during NATO's bombings. The body did not remain passive. The disciplining effect of the bombings was severe in that the NATO strikes both coincided with and assisted the Milosevic regime's intensification of its own disciplining of the bodies of the population of Belgrade and Serbia. This is illustrated in articles in newspapers such as the Serbian state-run *Politika* (*Politics*), which provided instruction on how to recognize and respond to different emergency sounds as well as patriotic texts on the vitality and nutritional value of national dishes, affirming that food was plentiful. In May, spurred by sporadic disruption of electricity and water supplies, *Politika*'s articles took on a more serious tone, advising on how to preserve foods and suggesting that 50 litres of water per day was adequate for all of an individual's cooking and hygiene needs. These reached their height in the leaflets, distributed to every household, addressing aspects of civil protection under the banner Urban Population Self-Protection Reminder (*Civilna zastita: podsetnik za samozastitu gradskog stanovnistva*).

The existence of the individual was under constant threat. However, the use of the body during the Songs Sustained Us gatherings exposed both the strength and fragility of the human body when deployed as a tool of resistance. This is evidenced in the human bodies that acted as live shields to protect Belgrade's Branko's Bridge, despite 60 road and rail bridges and overpasses in the Former Republic of Yugoslavia (FRY) already having been destroyed during the operation and the threat from NATO's Air General Michael Short on 15 May 1999 that included the words:

> I think no power to your refrigerator, no gas to your stove, you can't get to work because the bridge is down – the bridge on which you held your rock concerts and you all stood with targets on your heads. That needs to disappear.[89]

The importance of the body is the ultimate terrain of resistance where individuals, by inserting their bodies into a very politicized and contested space, became possible targets of NATO's strikes, whilst also becoming an open display of an individual's will to life. The deployment of the body as a form of resistance showed that the Operation's use of force and control could be challenged; history was not only activated but also power was reborn. This in itself shows the potential of Balkanism and balkanization when deployed for purposes of extending diversity and utilizing the excess and the abnormal; the events revealed an alternative reinhabitation and reinterpretation of public spaces; they not only moved beyond control societies, but also demonstrated the possibility of countering control societies through deployment of bodies in space.

Film

The Songs Sustained Us gatherings evoked the 'insanity' of Emir Kusturica's carnivalesque films in which the joy for life is always present. In Kusturica's *Underground* (*Podzemlje*, 1995), events of WWII and the violence of the 1990s onto which they are mapped are interchangeably brought into play and experienced by a community of people living underground and on the surface; the underground is symbolic of all that was once removed from the landscape of visibility. The film's themes of secrecy, gore, betrayal and, most important, nationalism, are not explored for purposes of reinforcing all these associations of violence in the former Yugoslavia; they are used for purposes of both purifying and compounding the constructs about the Balkans. This is done through the deployment of the very stereotypes of the Balkans as a feral zone of rogue violence and long-standing ethnic hatreds. The strategy here is to eclipse negativity through the negative imagery by which the construct has been made possible. Kusturica, in many ways, revels in the differences found in this region, from drinking in excess to partying and violence. Life resides recklessly in the Balkans; life and death, peace and war coexist in more transparent ways and are more passionately felt and disclosed.

However, no sooner does Kusturica set up these themes in *Underground* than he immediately pulls them down, suggesting that only those who have lived in the context can comprehend the situation in the former Yugoslavia. From this perspective, he both inverts the dichotomy of exterior and interior borders and confronts the slippery boundary of fact and fiction, reality and dream. The film's first scenes are a chaotic mix of sex, bombs and animals escaping from the zoo; these are soon interwoven with scenes from a film about WWII and the struggle of Partisans against the Germans. Yet one is never quite sure whether the film is just a film or an actual narration of real-life events. It is a simulation of events and times sliding past each other, and simultaneously brought into and out of each other.

His film enters a different time frame to the one found in Europe where events are set out along a chronological line. If Balkanism sits outside the history of civil Europe and the West, and if Bauman's position on the history of time beginning with modernity – whereby "modernity is the time when time has a history"[90] – is valid, then it is not a question of Balkanism lagging in time by not entering the Western history of Enlightenment (which Foucault identifies with various modes of classification and disciplining of every aspect of life), but simply that Balkanism exists outside the disciplinary codes of Europe. The film ends with a parcel of land splitting from the main territory, and thus the presence of Balkanism and balkanization remains bound and in continuous existence. The Balkans, in this instance, do not evoke Todorova's conception of this zone as reminiscent of the periphery and as such a space that is ghettoized in terms of its developmental processes.[91] Rather, it is that the newly balkanized territory of Kusturica's film shows that the Balkans and Balkanism are in their own right an alternative entity and force. To balkanize, using the Yugoslav context as a precedent, is done for purposes not only of creating a territory of diversity but also of sidestepping the binary 'us' and 'them' construct.

The symbolism of the title *Underground* is not only suggestive of the partial nature of perception of reality. It also proposes that the knowledge of violence is limited. The film's imagery of underground passages, used by the West as well as Yugoslav forces in the lead-up to and during the 1990s Yugoslav wars for purposes of shelter as well as for transport and storage of goods from food to aid supplies and munitions, presents a multilayered and perhaps unexplored narrative of violence within this context, violence that has remained largely absent from the mass media. The absence of violence is the physical presence of the underground in the film. The construction of a stable understanding of Balkanist violence and the wars that have been fought in this region challenges and resists the omnipresent belief that Balkanist history can be mapped in a straightforward manner. The underground tunnels make it possible to draw a parallel and argue that the hegemonic attempt to identify peace and civility with one side, and war and vulgarity with the other side, is not only unstable but also porous and can be challenged.

To compare the Balkans with Europe is simply to set up a dichotomy between a whole and its counterpart, allowing Europe's image to remain perennially untarnished and progressive, as it can project its threats through the absurdity of the peripheral other. This is to an extent the story line of the US film *Cat People* (1942, dir. Jacques Tourneur), in which an American average-Joe marries Irena Dubrovna, a Serbian artist. Their marriage is tainted by Irena's psychotic fear that she suffers from an ancient curse under which intimacy and arousal will turn her into a panther. Enlightened order is thus symbolically paired with primitive anxieties. This figure of the cat 'oddly' reappears in Kusturica's comedy *Black Cat, White Cat* (*Bela Macka, Crni Macor*, 1998). In his film – whose narration slips between Serbian, Bulgarian and Romani – the recurring motifs of black cat and white cat assist in framing questions to do with love, loyalty and friendship.

The understanding of elementary values is taken further with the setting of the film itself, located along the border of Serbia and Bulgaria; a zone where Balkanist passions reach great levels from matters to do with money, to feelings of joy and love and experiences of trauma. The two dichotomous cats, meanwhile, take on an all-knowing role; they are the only witnesses to events in the film not seen by others, they see beyond the immediate. Yet they are also the only witnesses of the true love between young Zare and Ida. Seemingly, there is something about the symbolism of the cat(s) that suggests that time and space in the Balkans evade all rational and linear thinking about them, and that values of love and loyalty both still pertain if perceived through less rational thinking. The operation of balkanization in the film is symbolically suggested through the presence of both a black and a white cat, and the ability of these not only to coexist but also be stronger in their perception and knowledge; this coexistence was seen in the former Yugoslav context where diversity was implemented when grouping together territories and people.

While Balkanism is conceptually open-ended, it also operates as a transitional spatial force that cannot quite fit any given mould. This may be because the behaviour patterns seen as normal in the 'civilized' world do not hold sway here.[92] Operationally, the potential of Balkanism's variable liminality is not necessarily about the marginality of this zone, if marginality gives this zone a sense of centrality,[93] but that this liminality opens up the potential of thinking identity outside the clear and concrete boundaries of a highly engineered society. Variable liminality (which is inclusive of being limit-distinct) evokes the thinking of Deleuze and Guattari, where identity is

> not the point where one no longer says I, but the point where it is no longer of any importance whether one says I. We are no longer ourselves. Each will know his own. We have been aided, inspired, multiplied.[94]

Thinking identity across the trajectory of liminality resists identification of Balkanism in terms of a binary logic (Christianity/Islam, civilization/barbarism, etc.). Instead, it facilitates thinking identity in terms of instability and complexity. To have instability is to accommodate a possibility to move time and history on from a state of perpetual replay in the cause of spreading Western normality and civil values by eliminating all that is considered abnormal and barbaric. What instability offers is the conception of identity as ever-changing and ever-arising; one that is many and where the origin is not only never given but also always remade.

Architecture and the urban

When considered architecturally, instability implies that both the ground and the horizon are dynamically fluctuating. The presence of dynamism was seen in Nikola Dobrovic's Generalstab complex design that emerged as the winner of the

1953 closed competition organized by the Yugoslav People's Army (*Jugoslovenska Narodna Armija* – JNA). It was the first architectural signifier of SFRY's 1948 split from Soviet Russia. Dobrovic approached the Generalstab complex not only as the means to deal with the requirements of the project on an urban and architectural level, but also as a formative resolution and demonstration of a future Grand Belgrade and, indirectly, Yugoslavia.[95] The project was imagined to showcase the construction of an alternative identity by enshrining the rights of different classes and nations through the concept of Yugoslavia as a federation.

According to Franke Wilmer, the new post-WWII Yugoslav identity was enveloped in several agendas: anti-imperialism, partisan heroism and the formation of an alternative socialist programme that attempted to sidestep the excesses and extremes of both Western capitalism and Soviet-style communism.[96] Against its symbolic identifications, Dobrovic's drive to assert a new dynamic urban vision for Belgrade through the design of the complex was supported by referencing the city's immediate surroundings; the city's shifting horizons facilitated by flat horizontal ground and vertical topographical points broken and brought together by the rivers Sava and Danube.[97] His vision attempted to rethink and reinterpret space and society beyond the framework devised by the state or dominant Western modernist ideology.

The notion of dynamism seems to have arisen out of Dobrovic's three specific interests. The first was his interest in philosophy; particularly Henri Bergson's ideas on matter and memory. The second was his interest in art. The third was the necessity he saw to reinterpret CIAM (The International Congresses of Modern Architecture)[98] to suit the context of Belgrade and Yugoslavia. All of this culminated in the idea of dynamism, which was orchestrated in Dobrovic's design of a narrow volume measuring 250m in length. The volume was set back from the street and stretched across from one end of the site to the other. This move not only facilitated a 270m-wide field to spatially orchestrate his vision of space as a series of shifting horizons, but also allowed breathing space for the surrounding historical buildings.[99]

The void within the complex, made possible due to the interval of space between its two buildings, itself split by a road, offered the symbolic contextualization and the memory of the Sutjeska Offensive; the fatefulness of the battle for the future of Yugoslavia was brought to the capital city of Belgrade.[100] This is significant since in the West following the end of WWII there was an attempt to delete memory from architecture.[101] Dobrovic's drawing out of a relation to the Sutjeska Offensive in 1960 brought together the thinking behind Henri Bergson's 'cone' with a historical event.[102] The relation between them was that an event transforms identity in time through becoming; in this instance, with the formation of Yugoslavia as an alternative brand of Socialism (Titoism) that was not aligned with the agenda of Cold War politics.[103] In terms of the actual 1943 battle, the memory of the event is wrapped in the narrative of 120,000 Germans failing to break the Partisan formations of less than 20,000 soldiers or capture then commander Josip Broz Tito. The battle in southeastern Bosnia and Herzegovina lasted for a whole month and was directly vested in realigning the morale, position and unfolding of the events to come in breaking the Fascist grip. Considering the mismatch in numbers of soldiers on the two sides, it is unsurprising that this event took on epic proportions in the new narrative of SFRY identity; a narrative framed by bravery, strength and will.

According to Vladimir Kulić, were it not for the symbolism of Sutjeska as the key element of the Generalstab complex, the void would have no meaning.[104] However, the more important focus, and one that largely remains unexamined, is not whether the void within the complex was a literal insertion of an epic offensive that occurred in the 'V'-shaped Sutjeska valley. It is rather that the void is a signifier that the event was transformative not only of that particular moment but also of memory and history, since history was opened up by the event. History as a preconditioned state of affairs was altered through an act of insertion in time. The event facilitated the more affirmative and heterogeneous balkanization of the second Yugoslavia; balkanization that arose through extending the heterogeneity of values and rights. The role of architecture in this process is that it accommodated this alternative identity by reinterpreting CIAM modernism through the void; signified by the unpredictable movement of the offensive and the dynamism of Bergson's cone which undermine cartesian flattening of identity and perception. Thus, the conception of a Balkanist identity is not in the delineation of either and/or, but in the overlapping of at least two spaces, for example, the multilayered and intersecting historical, geopolitical and socioeconomic context of Belgrade. One can speculate that this identity may be fluctuating and open, and in this state challenges the conventional notion of borders being binary and fixed.

For Bergson, the relation between memory and perception is found along two axes. The first axis is based on rotation, where memories come together and converge from the 'present' moment; such memory is of a mechanical and habitual nature. The second axis is based on concentration of more complex perceptions, such as those made available to us through dreams; memory here is of cosmic nature in that it extends beyond individual memory, it is encompassing of memory beyond the individual. Within this configuration based on elastic expansion-contraction-rotation, one is no longer a mere observer of life but an active creator, since identity and life are constituted through multiple layers of memory in extension, with no layer having a clear beginning or end.[105] The significance of Bergson's philosophy to Dobrovic's architecture was seen in the architect's initial four-option proposal for the design of the complex which he called "Bergson dynamic schemes"[106] as well as in Dobrovic's assertion that Bergson's conceptualizations are not a matter of idealism but a base upon which modern architecture and its theory may be conceived.[107] It is not only that Dobrovic brought aspects of the Sutjeska Offensive to Belgrade in the form of a void, but that he also offered an opportunity to remake the identity of that site and make transformation an active part of Belgrade and the new Yugoslav identity. While he had a vision for the new formation, Dobrovic felt strongly that individual visions required translation into something solid and actual, with a social imperative. While a dose of abstraction and ambiguity are required, when visions remain at the level of visions, architecture suffers.[108]

It is not just that the transformative nature of the Sutjeska Offensive has remained unmentioned since NATO's 1999 intervention, but previous associations with the Generalstab complex have become obscured beneath processes of its erosion and demolition, both of which are seemingly contingent on negation and forgetting.[109] The Generalstab complex was targeted twice, on the nights of 29 April and 30 April and of 7 May and 8 May. These two nights saw the most forceful strikes in terms of the overall urban infrastructure damage caused to Belgrade. The delay in the targeting of this complex, despite its strategic importance, is perplexing, considering NATO's initial strategy was presented as being proportional, meaning that only military and police institutions associated with the Slobodan Milosevic regime were to be targeted. Since it is likely that the 1991 urban destruction of Vukovar, the 1991 bombings of Dubrovnik and the 1992 siege of Sarajevo were all ordered from Belgrade's Generalstab complex,[110] the complex was predictably a prime NATO target. Yet one of the first targets of NATO's 78-day campaign was an empty factory just outside the city. During that same night, three empty schools as well as a nearby monastery in suburban Rakovica[111] were partially destroyed. A more relevant and 'justifiable' target on that first night was the military air base in the Belgrade suburb of Batajnica.

Figure 2.5 The filleted Generalstab complex in 2010, with the Military Headquarters (Building A) on the left-hand side incurring greater damage than the Ministry of Defence (Building B).

Immediately after the targeting, the complex (un)intentionally became a memorial and a cultural artefact associated with Operation Allied Force. More recently, the role of the complex in its past, present and future has been reduced to eliminating any memories that deviate from the somewhat corrupt politics of control found within the Property Directorate of the Serbian Government and the Serbian Army. The lack of awareness regarding the image of the Sutjeska Offensive as an event that affirms life and facilitates a creative leap in history is a signifier of Serbia's current passive and stagnant identity. Ironically, the stagnation has been reinforced through the political changes post-1999: the October 2000 democratic elections; rapid privatization of land, property and infrastructure; and the introduction of a consumer culture. A new 'democratic' government also meant a shift in the country's policy towards integration with Europe, including a shift on questions related to security measures following the script determined by Europe, that is, NATO.

The changes are geared towards globally eliminating any difference in political opinion; a new type of Iron Curtain is being drawn. It is framed by the push to absorb all political difference of Balkanism and balkanization in terms of grouping together and accommodating diversity; the changes are directed towards establishing a perpetual present that eliminates any rifts, meaning that the individual and social are the ever-passive audience; to live and to think is to observe time as a condition of habit, rather than as a condition of movement between contemplation, action and change. The latter is suggestive of Bergson's thinking on matter and memory in relation to the cone; pure memory found on a plane at the base of the cone moving forward into singular images.

The concentration of imagery was directed towards action and creation during the conception and actualization of the Generalstab complex, which was completed despite Dobrovic parting ways with the project due to differences of opinion with the JNA. The West initially saw sufficient value in the project to merit providing funding for the Generalstab complex by way of a World Bank grant, at a time when buttressing of Yugoslavia as a viable political force not aligned with the Cold War politics of Western Europe or Soviet Russia was of importance in establishing and strengthening Belgrade as a cultural, architectural and political centre. However, when the building was finished, it was perceived by the West as an architectural nightmare of concrete and pink stone façade, and an indicator of Communism.[112] This shift in perception is significant, as from the end of WWII to the completion of the Generalstab complex in the mid-1960s, Yugoslav architecture was built mainly in the modernist International style and, unlike that of other communist countries, was praised by the West.[113]

It was only when Yugoslavia started to clearly form its own alternative identity, signified by the Generalstab complex, that Yugoslav architecture came to be perceived in the West as an architectural nightmare. The nightmare associated with the complex was attributed to Dobrovic's integration of modernist architecture, philosophy and local materials (red stone came from the Kosjeric Plant in Serbia, and white marble from Brac, an island in Croatia), which might be claimed to prefigure the later arrival of postmodernism in the West. In a lecture given on 31 January 1963 in the amphitheatre of the Faculty of Architecture in Belgrade, and at the invitation of the Society of Architects and Technicians of the Socialist Republic of Serbia, Dobrovic said:

> What is our task in this part of the globe? Is it not about grinding those materials that have been brought here, with no customs inspection, in the domestic mill and while grinding combining them with all those domestic additives and spices, domestic brains and hearts, which will turn it into our product, our national school's product?[114]

The fact that in the 1960s the SFRY was starting to reinstate political and military communication with the Soviets did nothing to enhance the perception of Yugoslav architecture in the West. Up until then, the SFRY was presented in a politically positive light due to the 1948 split from the USSR, which the West interpreted as a hopeful sign that other socialist and communist satellite states would follow and, in turn, undermine the USSR's power. More recently, there has been a strange subversion of all that arose and prevailed during the period of Socialist Yugoslavia; it is largely portrayed in terms of mere nostalgia, backwardness rather than deliberation, and as that which stood in the way of the lifestyle of choice and freedom associated with the English-speaking Western world.

Seemingly, whether the spectre of Balkanist violence is invoked is contingent on alliance and integration with the West. That Yugoslav political orientation remained unclear and, as such, unpredictable, would have also been cause for concern. In a 1977 lecture, "Global Balkanization," the Russian-American novelist/philosopher/screenwriter Ayn Rand associated the ways of living found in areas ranging from Catalonia and the Basque region all the way to Yugoslavia with separatism and pre-civilized ways of thinking statehood.[115] Particular to Yugoslavia, balkanization was seen as a tendency to endless warring against other ethnic tribes. Along similar lines, balkanization in this configuration was critiqued as incompatible with the preferred Anglo-American capitalism, due to its abnormal and odd ways of existence.[116] To cooperate and align with the West from the 1970s was to embrace globalization of the market economy, that is, privatization, consumerism and surveillance. The agenda of global politics was certainly present in Yugoslavia; however, its implementation was based on charting a path that diverged from the Western one.

This alternative way of practicing balkanization and globalization was seen during Yugoslavia's period of non-alignment from the 1960s.[117] On an urban and architectural level, firms such as Energoprojekt (Energyproject), Rad (Work), Mostogradnja (Building Bridges) and Komgrap from Belgrade, Energoinvest (Energyinvest) and Hidrogradnja (Hydrobuild) from Sarajevo and INGRA and Rade Koncar from Zagreb were guarantors for projects covering architecture and infrastructure systems in Africa, Asia and South America: Libya, Nigeria, Zambia, Zimbabwe and Nigeria, to Iraq, Pakistan and Kuwait, and Peru. Each firm had the skill and capability to oversee every aspect of design and construction and transfer this knowledge globally. The motive behind Tito's globalizing strategy was to provide a quality product at a lower cost than Western competitors, who not only charged more but also were not members of the Non-Aligned Movement (NAM). That self-management was inbuilt into the Yugoslav system meant that any profits won for early project completion were also distributed amongst the individuals involved in the particular project. Globalization was a question of evening out socioeconomic disparities, rather than amplifying them, as seems to be the norm today. Thus, NAM was a project of alternative globalization, globalization from the periphery.

In terms of the Generalstab complex, the evening out of disparities was by conceptualizing space as an opportunity to accommodate events and think perception as dynamically shifting frames. Dobrovic saw it as important to rethink ideas and the role of architecture; he compared static and traditionally comprehended architecture with "accidentally fallen dust."[118] However, his thinking has an inversion: that of accidental fallen dust as a discharge and a spatial strategy to reimagine alternative futures whose projected identity is not predetermined or fixed, it is *under construction*. In other words, within the extended void facilitated by NATO's targeting, the fallen dust of the complex has not only destabilized the structure and a conceptualization of what a void may be but has also put into question whether decay and dust historically come from the inside out or outside in, which, in turn, marginalizes and reframes structures of power. The impasse of identification is the power of Balkanism.

Notes

1 Some notable authors include Maria Todorova, Vesna Goldsworthy, Milica Bakic-Hayden and Robert Hayden, Dušan I. Bjelić, Obrad Savić, Rastko Močnik, Aleš Debeljak, Slobodan G. Markovich, Tomislav Longinovic, Stathis Gourgouris, Ugo Vlaisavljević, Ludmilla Kostova, Christina Koulouri, and Stjepan G. Meštrović.
2 Dušan I. Bjelić, "Introduction: Blowing Up the 'Bridge,'" in *Balkan as Metaphor*, ed. Dušan I. Bjelić and Obrad Savić (Cambridge, MA: MIT Press, 2002), 4.
3 Mark Mazower, *The Balkans* (London: Phoenix Press, 2001), 1–2. Interestingly, the Balkan mountain range was seen to stretch from Black Sea to the Adriatic, rather than its actual location from Bulgaria to the river Timok in eastern Serbia.
4 Frederick Calvert Baron Baltimore, *A Tour to the East, in the Years 1763 and 1764. With Remarks on the City of Constantinople and the Turks* (London: W. Richardson and S. Clark, 1767).
5 Mazower, *The Balkans*, 1–16.
6 Andrew Baruch Wachtel, *The Balkans in World History* (New York: Oxford University Press, 2008), 8.
7 Wachtel, *The Balkans in World History*, 72–73.
8 Mazower, *The Balkans*, 77; Todorova, *Imagining the Balkans*, 3.
9 H. T. Greenwall, "Head of Great Industry Predicts the 'Balkanization of Europe,'" *The New York Times*, 20 December 1918.
10 Greenwall, "Head of Great Industry Predicts the 'Balkanization of Europe.'"
11 Paul Scott Mowrer, *Balkanized Europe: A Study in Political Analysis and Reconstruction [1921]* (Ithaca: Cornell University Press, 2009).
12 Mowrer, *Balkanized Europe*, 160.
13 Mowrer, *Balkanized Europe*, 279.

14 Eric J. Hobsbawm, *Nations and Nationalism Since 1780: Programme, Myth, Reality* (Cambridge: Cambridge University Press, 1990), 31–32.
15 Todorova, *Imagining the Balkans*, 33.
16 Robert D. Kaplan, "In the Balkans, No Wars Are 'Local,'" *New York Times*, 7 April 1999, cited in Kaplan, *Balkan Ghosts: A Journey Through History* (London: Picador, 2005), xxviii.
17 Andrew Hammond, "Balkanism in Political Context: From the Ottoman Empire to the EU," *Westminster Papers in Communication and Culture* 3:3 (2006): 19.
18 Todorova, *Imagining the Balkans*.
19 Maria Todorova, "The Balkans: From Discovery to Invention," *Slavic Review* 53:2 (1994): 453–482; Wachtel, *The Balkans in World History*, 72–73.
20 Vesna Goldsworthy, "Not Quite European . . . Scrutinizing the Balkan Neo-Colonialism," *Eurozine* (2004).
21 Todorova, *Imagining the Balkans*, 16.
22 Interview with Aleš Debeljak, "On the Balkans," *Balkans Project*.
23 Alexander Kiossev, "The Dark Intimacy: Maps, Identities, Acts of Identification," in *Balkan as Metaphor*, ed. Dušan I. Bjelić and Obrad Savić (Cambridge, MA: MIT Press, 2002), 165–191.
24 Goldsworthy, *Inventing Ruritania*.
25 Goldsworthy, *Inventing Ruritania*.
26 Bjelić, "Introduction: Blowing Up the 'Bridge,'" 11.
27 Goldsworthy, *Inventing Ruritania*.
28 Todorova, *Imagining the Balkans*, 12.
29 Elissa Helms, "East and West Kiss: Gender, Orientalism, and Balkanism in Muslim-Majority Bosnia-Herzegovina," *Slavic Review* 67:1 (Spring 2008): 90.
30 Rastko Močnik, "The Balkans as an Element in Ideological Mechanisms," in *Balkan as Metaphor*, ed. Dušan I. Bjelić and Obrad Savić (Cambridge, MA: MIT Press, 2002), 96.
31 Milica Bakic-Hayden and Robert Hayden, "Orientalist Variations on the Theme 'Balkans': Symbolic Geography in Recent Yugoslav Cultural Politics," *Slavic Review* 51:1 (1992): 8; Milica Bakic-Hayden, "Nesting Orientalisms: The Case of the Former Yugoslavia," *Slavic Review* 54:4 (1995): 920.
32 Bakic-Hayden and Hayden, "Orientalist Variations," 4.
33 Edward W. Said, *Orientalism* (London: Penguin Books, 2003).
34 Močnik, "The Balkans as an Element in Ideological Mechanisms," 79–119.
35 Todorova, *Imagining the Balkans*, 194.
36 Andrew Wear, *Knowledge and Practice in English Medicine, 1550–1680* (Cambridge: Cambridge University Press, 2000).
37 Amy Swiffen, *Law, Ethics and the Biopolitical* (New York: Routledge, 2010).
38 Cited in Todorova, *Imagining the Balkans*, 35.
39 Todorova, *Imagining the Balkans*, 35.
40 W. E. B. Du Bois, *The World and Africa: Colour and Democracy, Colour and Race* (New York: Oxford University Press, 2007), 72–73.
41 D. Parthasarathy, "Decentralization, Pluralization, Balkanization? Challenges for Disaster Mitigation and Governance in Mumbai," *Habitat International* 52 (2016): 31.
42 Octavia E. Butler, *Parable of the Sower* (New York: Four Wall Eight Windows, 1993).
43 Michael Dear, "Los Angeles and the Chicago School: Invitation to a Debate," *City & Community* 1:1 (2002): 18. See also: Alberto Alesina and Eliana La Ferrara, "Participation in Heterogeneous Communities," *Quarterly Journal of Economics* 115:3 (2000): 847–904.
44 Mike Davis, "Fortress Los Angeles: The Militarization of Urban Space," in *Variations on Theme Park*, ed. Michael Sorkin (New York: Hill and Wang, 1992), 155.
45 Douglas S. Massey and Nancy N. Denton, *America Apartheid* (Cambridge, MA: Harvard University Press, 1993).

46 Peter T. Leeson, "Balkanization and Assimilation: Examining the Effects of State-Created Homogeneity," *Review of Social Economy* LXV:2 (2007):141. See also: Alejandro Portes and Robert L. Back, "Immigrant Earnings: Cuban and Mexican Immigrants in the United States," *International Migration Review* 14 (Fall 1980): 315–341; Alejandro Portes and Alex Stepick, "Unwelcome Immigrants: The Labor Market Experiences of 1980 (Mariel) Cuban and Haitian Refugees in South Florida," *American Sociological Review* 50 (August 1985): 493–514; Loic Wacquiant, *Urban Outcasts* (Cambridge: Polity, 2007).

47 See: Neil Brenner and Roger Keil, eds., *The Global Cities Reader* (New York: Routledge, 2005); Howard E. Aldrich and Roger Waldinger, "Ethnicity and Entrepreneurship," *Annual Review of Sociology* 16 (August 1990): 111–135.

48 See: John R. Logan, Richard D. Alba and Thomas L McNulty, "Ethnic Economies in Metropolitan Regions: Miami and Beyond," *Social Forces* 72:3 (March 1994): 691–724; Portes and Back, "Immigrant Earnings"; Portes and Stepick, "Unwelcome Immigrants."

49 Bill Clinton, "Clinton Rodney King Trial Remarks," *C-SPAN.org*, 17 April 1993 (user-created clip, 19 July 1993), <www.c-span.org/video/?c4459701/clinton-rodney-king-trial-remarks>.

50 D. G. Shane, "Balkanization and the Postmodern City," in *Mortal City*, ed. Peter Land (New York: Princeton Architectural Press), 56. The term heterotopias was used by Michel Foucault in his 1967 paper *Of Other Spaces*, which remained unpublished until 1984. It was written with the intent of being delivered only in a lecture to a group of architecture students. *Of Other Spaces* introduces heterotopias as spaces outside everyday socially constructed and disciplining norms. Foucault borrowed this term from medical terminology, where heterotopia was used to describe tissue conditions that appeared abnormal in respect to the overall structure. In medical terminology, heterotopias appear to have a marginal status and do not prevent the overall organism from functioning, while Foucault's heterotopias are considered as spaces essential to the city, and as counter sites. While some heterotopias have a disciplinary structure, such as those found in prisons, psychiatric hospitals and military institutions, others such as theatres and ships are more open and transformative. For more information, refer to Michel Foucault, "Of Other Spaces (1967)," in *Heterotopia and the City: Public Space in a Postcivil Society*, ed. Michiel Dehanene and Lieven De Cauter (London: Routledge, 2008), 13–29.

51 Shane, "Balkanization and the Postmodern City," 56.

52 Shane, "Balkanization and the Postmodern City," 58–64.

53 Saskia Sassen, "Europe's Migrations. The Numbers and the Passions Are Not New," *Third Text* 20:6 (November 2006): 635–645; Saskia Sassen, *Guests and Aliens* (New York: The New Press, 2000); Saskia Sassen, *Losing Control? Sovereignty in an Age of Globalization* (New York: Columbia University Press, 1996).

54 Aristide R. Zolberg and Long Litt Woon, "Why Islam Is Like Spanish: Cultural Incorporation in Europe and the United States," *Politics and Society* 27:1 (March 1999): 5–38.

55 Martin Heidegger, "Building Dwelling Thinking," in *Basic Writings* (London: Routledge, 2010), 348. Note: Heidegger's link to Nazism may have affected the understanding of the text and its linking to questions of ethnicity.

56 William Kurt Barth, *On Cultural Rights: The Equality of Nations and the Minority Legal Tradition* (Boston: Martinus Nijhoff, 2008).

57 James Goudkamp, *Tort Law Defences* (Oxford: Hart Publishing, 2013). Note: the examples tend to focus on law processes in Australia.

58 Jonathan H. Adler, "Climate Balkanization: Dormant Commerce and the Limits of State Energy Policy," *LSU Journal of Energy Law and Resources* 3:1 (Fall 2014): 159.

59 Adler, "Climate Balkanization," 193.

60 Angela Mitropoulos and Matthew Kiem, "Cross-Border Operations," *The New Inquiry*, 18 November 2015, <https://thenewinquiry.com/cross-border-operations/>.
61 Angela Mitropoulos, "Archipelago of Risk: Uncertainty, Borders and Migration Detention Systems," *New Formations* 84–85 (2015): 164.
62 Out of 68.5 million, 40 million are internally displaced people, 25.4 million are refugees and 3.1 million are asylum-seekers. Refer to UNHCR, "Figures at a Glance," 2018, <www.unhcr.org/figures-at-a-glance.html>.
63 Jonah Hill, "Internet Fragmentation: Highlighting the Major Technical, Governance and Diplomatic Challenges for U.S. Policy Makers," Report, Belfer Center for Science and International Affairs (BCSIA) – Harvard Kennedy School (HKS), 2012.
64 John D. Woodward, Jr., "The Law and the Use of Biometrics," in *Handbook of Biometrics*, ed. Anil K. Jain et al. (New York: Springer, 2007), 357–379.
65 Woodward Jr., "The Law and the Use of Biometrics."
66 The need to consider the EU's alternatives was noted by Angela Merkel at a Franco-German summit in Paris in February 2014.
67 Leslie Benson, *Yugoslavia: A Concise History* (Basingstoke: Palgrave Macmillan, 2004).
68 All photos in this book are author's own.
69 Prior to WWI, only Serbia and Montenegro were independent Kingdoms.
70 Benson, *Yugoslavia: A Concise History*, 51.
71 Ivan Žaknić, *Journey to the East* (Cambridge, MA: MIT Press, 2007), 43.
72 Žaknić, *Journey to the East*, 43.
73 Žaknić, *Journey to the East*, 43–49. Note: Le Corbusier was more complimentary about the Romani music as well as the items exhibited (carpets, pots and clothing) in the ethnographic museum in Belgrade. In 1965, he apologized for the negative descriptions of Belgrade, the explanation being that the perception may have been affected by his young age; he was only 23 years old.
74 The Council was established in November 1942 in Bihac – a city in northwestern Bosnia – to administer all areas controlled by the Partisans. The Partisans were a communist-led resistance to the Axis powers led by Marshal Josip Broz Tito.
75 Dejan Jović, *Yugoslavia: A State that Withered Away* (West Lafayette: Purdue University Press, 2009), 78.
76 Dorian Jano, "From 'Balkanization' to 'Europeanization': The Stages of Western Balkans Complex Transformations," *L'Europe en formation* 349–350 (2008): 55.
77 Pål Kolstø, " 'Western Balkans' as the New Balkans: Regional Names as Tools for Stigmatisation and Exclusion," *Europe-Asia Studies* 68:7 (2016): 1246; Andrew Hammond, *The Balkans and the West* (London: Routledge, 2004), xiv.
78 Todorova, *Imagining the Balkans*, 191.
79 Croatia is 99 per cent white, and over 90 per cent of its population is Croatian Catholic.
80 It is true that the question may arise as to whether Germany's admission to the EU has also depended on convenient forgetting. However, accommodation of Other histories and acknowledgement of genocide has occurred nationally there, and the effects are manifest in education as well as arts and culture such as films and architectural memorials.
81 Paul Patton, "Conceptual Politics and the War Machine in Mille Plateaux," *SubStance* 13:3/4 (1984): 63.
82 Christine Lavrence, " 'The Serbian Bastille': Memory, Agency, and Monumental Public Space in Belgrade," *Space and Culture* 8:1 (2005): 32.
83 Gilles Deleuze, *The Logic of Sense* (New York: Columbia University Press, 1990), 1, 3.
84 Foucault, "Of Other Spaces (1967)," 22.
85 Thomas L. Friedman also served as chief economic correspondent in the newspaper's Washington bureau. Prior to that, he was the chief White House correspondent. In 2005, Friedman was elected as a member of the Pulitzer Prize Board.

86 Thomas L. Friedman, "Foreign Affairs; Stop the Music," *New York Times*, 23 April 1999.
87 Stephen Graham, "Disruption by Design: Urban Infrastructure and Political Violence," in *Disrupted Cities: When Infrastructure Fails*, ed. Stephen Graham (New York: Routledge, 2010), 116.
88 Samuel Weber, "War, Terrorism, and Spectacle, or: On Towers and Caves," *Grey Room* 7 (2002): 14–23.
89 Cited in Michael Almond, "Belgrade Open for Business as Usual," *The Guardian*, 16 May 1999.
90 Zygmunt Bauman, *Liquid Modernity* (Cambridge: Polity Press, 2000), 110–112.
91 Maria Todorova, "The Trap of Backwardness: Modernity, Temporality, and the Study of Eastern European Nationalism," *Slavic Review* 64:1 (2005): 140.
92 Todorova, *Imagining the Balkans*, 3.
93 Katherine E. Fleming, "Orientalism, the Balkans, and Balkan Historiography," *The American Historical Review* 105:4 (2000): 1218–1233.
94 Deleuze and Guattari, *A Thousand Plateaus: Capitalism and Schizophrenia*, 3.
95 Marta Vukotić Lazar, *Beogradsko Razdoblje Arhitekte Nikole Dobrovića (1945–1967)* (Beograd: Plato, 2002), 64.
96 Franke Wilmer, *The Social Construction of Man, the State, and War: Identity, Conflict and Violence in Former Yugoslavia* (New York: Routledge, 2002), 88.
97 Lazar, *Beogradsko Razdoblje Arhitekte Nikole Dobrovića (1945–1967)*.
98 The International Congresses of Modern Architecture (CIAM) was founded in Switzerland in 1928 with the aim of using the fields of architecture – from landscape architecture to urbanism – to spread the principles of the Modernist Architecture socially, economically and politically. Some of CIAM members included Le Corbusier, Sigfried Giedion, Gerrit Rietveld, Walter Groupius and Alvar Aalto. The demise of the organization occurred in 1959.
99 For further detail see Nikolina Bobic, "Belgrade in Formation(s): Dobrovic's Generalstab Complex," *Fabrications* 21:1 (2012): 7–28.
100 Slobodan Milošević, "The Ghosts of the Past, Present and Future: The Case of the Army Headquarters in Belgrade, Serbia," *Journal of Architecture and Urbanism* 39–1 (2015): 37–55.
101 Adrian Forty et al., eds., *The Art of Forgetting* (Oxford: Berg Publishers, 2001); Efrat Ben-Ze-ev et al., eds., *Shadows of War: The Social History of Silence in the Twentieth Century* (Cambridge: Cambridge University Press, 2010).
102 Miloš R. Perović and Spasoje Krunić, eds., *Nikola Dobrović: Eseji, Projekti, Kritike* (Beograd: Arhitektonski fakultet Univerziteta, Muzej arhitekture, 1998), 331.
103 Titoism was formed after Tito parted with the USSR in 1948. Those Yugoslavs who were openly in disagreement with Tito's decision were sent to the political prison on Goli Otok (Barren Island). Emir Kusturica addressed this thematic in his 1985 film *When Father Was Away on Business*.
104 Vladimir Kulić, "Prostor u kretanju,' *Republika* (2010).
105 Henri Bergson, *Matter and Memory* (New York: Zone Books, 2005).
106 The first of these 'void tensions' was symmetrical; the second asymmetrical; and the last two extravagant. The simpler symmetrical mode was chosen (and built) due to financial constraints. The 'void tensions' literally evoked Bergson's 'cone.'
107 Nikola Dobrović, *Savremena arhitektura 3* (Beograd: Zavod za izdavanje udžbenika Socijalističke Republike Srbije, 1963), 59.
108 Dobrović, *Savremena arhitektura 3*, 232.[
109 In 1961, Dobrovic received the highest Yugoslav award for architecture, the October Award, for the smaller Building A – the Military Headquarters – which was completed under Dobrovic's direction in 1961. In 2007, the Army and the Ministry of Defence put forward a proposition to remove the historical preservation order so the buildings could be demolished and the land sold to a private investor.

110 Srdjan Jovanović Weiss, "NATO as Architectural Critic," *Cabinet* 1 (2000/01), <www.cabinetmagazine.org/issues/1/NATO.php>.
111 William M. Arkin, "Smart Bombs, Dumb Targeting?" *Bulletin of the Atomic Scientists* 56:3 (2000): 47–48.
112 Vladimir Kulić, "East? West? Or Both? Foreign Perceptions of Architecture in Socialist Yugoslavia," *The Journal of Architecture* 14:1 (2009): 137.
113 Kulić, "East? West? Or Both?" 130–137.
114 Lazar, *Beogradsko Razdoblje Arhitekte Nikole Dobrovića (1945–1967)*, 218.
115 Ayn Rand, *Global Balkanization* (Alviso: Palo Alto, 1977), 3–15.
116 Rand, *Global Balkanization*, 3–15.
117 Following the SFRY split from Soviet Russia in 1948, the SFRY became a part of the Non-Aligned Movement (NAM) in 1961. The NAM was conceived as a network of nations that had a 'third way' political geography and offered an alternative to the Cold War politics of the USSR/Warsaw Pact and the US/NATO. Those at the forefront of the NAM, who were also its conceptual initiators, were the presidents of three countries: India's Jawaharlal Nehru as a mediator and moral force for peace; Egypt's Gamal Abdel Nasser, who was against colonialism; and the SFRY's Josip Broz Tito, a defender of national independence between the two blocks, establishing an independent and experimental brand of socialism called Titoism (communism and capitalism combined). When the NAM was formalized, two more presidents became synonymous with this Movement: Ghana's Kwame Nkrumah and Indonesia's Ahmed Sukarno.
118 Perović and Krunić, eds., *Nikola Dobrović*: *Eseji, Projekti, Kritike*, 71.

3 Yugoslavia in balkanization

Beyond civil war, beyond ethnonationalism

Dissent and war

Alternative thinking to the 1990s balkanization of Yugoslavia

Balkanization as a geopolitical parcelization of the SFRY through dissolution of its long-standing heterogeneity has been operational from the 1990s. The international rhetoric has it that this dissolution has been driven by ethnonationalism. Ethnonationalism, a milder form of racism, is a drive to achieve ethnic purity within the borders of a given territory. However, the alternative position to the commonly accepted one is that while nationalism was used as a tool of manipulation to instate and drive both the dissolution and the subsequent conflict, it was not the sole driver. Neither is the narrative that the war was a result of Balkanist hatreds present amongst the ethnicities in this region the whole story. The balkanization of Yugoslavia was a potent mix of economic recession, ideological control through nationalism and mobilization of mass media; the destruction of cities in Yugoslavia and proclamations regarding rights to land and territory started after the gaining of international support for national self-determination.

Against a diagnosis of nationalism, in July 1991, and several days after the war in Croatia began, over 150 intellectuals and activists from every part of Yugoslavia held a meeting in Belgrade to appeal to the international community to guarantee protected status for Bosnia and Herzegovina, which at the time was still peaceful.[1] That same year saw large peace demonstrations held in Belgrade, Sarajevo and Skopje.[2] In late September 1991, with the war in Croatia already underway, a peace march of 10,000 people spread across the Balkan region, with its final destination being Sarajevo. The march created a human chain connecting the (Croatian) Catholic Church, the (Serbian) Orthodox Church, the (Muslim) Mosque, and the (Jewish) Synagogue.[3] These events counter a common belief that no alternatives were put forward to the international community.[4] In other words, while nationalism was certainly present and a notable factor in the dissolution of the SFRY, it is problematic when nationalism is seen as the sole factor.

For such a complex dissolution, the term ethnic conflict has been used uncritically to describe the war and fragmentation of SFRY, as if there were something self-evident and organic rather than constructed about that category.[5] The surge of attention to and understanding of those events was fed by an uninterrupted circulation of (often) unverified references, storylines and discourses. The category thus inscribes onto the whole nation a specific behaviour, which, in turn, is used to validate the Balkanist imaginary. To single out the Yugoslav dissolution as a unique instance of conflict that arose out of deeply seated Balkanist hatreds, nationalism and differing cultural values is to flatten the underlying motives and significance of wars such as WWII, the Cold War or the conflicts in Northern Ireland. If one were to characterize the dissolution of Yugoslavia as a signifier of ethnic brutality, then surely the nations that participated in those other wars would also risk being categorized as barbaric cultures.

From a different perspective, the belief that the dissolution was a result of ethnic hatreds is a colonial way of framing balkanization.[6] The current colonial framing to do with conflicts in this region is different from the conquests of the Ottoman Empire. The heterogeneity found in the Balkans was largely respected by the Ottoman Empire up until the 17th Century. Conversion to Islam – largely in Bosnia and Herzegovina – occurred for social status, the right to hold onto land and estates and to avoid paying higher taxes. The late 1970s and early 1980s economic crisis in Yugoslavia, exacerbated by the need to repay debt from post-WWII loans from the World Bank and the IMF, contributed both to an erosion of the socialist ideology of brotherhood and unity and to being used as a trigger to inflame nationalist tensions; the economic downturn in Croatia and Slovenia was politically promoted as a direct outcome of socialism.

The Western subsidies and loans from the late 1940s and early 1950s did initially facilitate a surge of productivity and economic stability in the following decades; but from the 1970s, the interest payments on the debt were increasing at an uncontrollable rate. A debt of US$4 billion in the early 1970s tripled by the end of the decade. Needless to say, the 1979 global oil crisis and recession hit Yugoslavia's precarious economic situation hard.[7] The pattern of economic downturn continued into the 1980s, with the 1983 debt recorded at US$20.5 billion and rising; it was the highest per capita debt in Eastern Europe. Having little in foreign exchange reserves, even everyday items such as coffee and petrol became scarce; the economic crisis was clearly spreading with no end in sight.[8] However, the concurrence of the economic crisis in Yugoslavia with the World Bank's and the IMF's 1980s economic liberalization only exacerbated the crisis which opened up Yugoslavia to privatization and market economics and the removal of public ownership and state control. The situation steadily deteriorated, reaching the climactic point in 1989 when inflation stood at 2,000 per cent and the unemployment rate at 15 per cent. The state of economic collapse was particularly evident in Kosovo, Macedonia and Montenegro, all of which declared bankruptcy in 1987. Bosnia and Herzegovina looked to be going down the same road with the collapse of significant industrial complexes such as Agrokomerc.[9]

In the context of economic depression, individual republics – particularly the most economically stable, such as Croatia and Slovenia, that were obliged to fund over 50 per cent of federal budget outlays – started to pursue policies that benefited themselves rather than the country as a whole. The economic gap between the wealthiest republic, Slovenia, and the poorest, Kosovo, tripled in the 1960s, and sextupled in the 1980s. Understandably, Croatia and Slovenia's position was that the socialist policy of economic equalization in the name of brotherhood and unity was not put into practice. Their preference was that if any redistribution were to occur it needed to happen directly from one republic to another rather than via the federal government in Belgrade. The belief was that the economic set-up privileged Serbia the most, since during the formalization of Yugoslavia post-WWII, this republic was exempted from the need to contribute towards the federal budget.[10]

The central government in Belgrade was undermined not only by the economic downturn, with its calamitous spike in unemployment rates, but also by the failure to find a new unifying voice to smooth over disagreements in the federation following the death of Tito in 1980. The dissolution that ensued was a result of economic and political fracturing, rather than purely of ethnic hatreds. By the end of 1991, Yugoslavia existed no longer; both Slovenia and Croatia had gained independence and adopted political systems that abandoned socialism. The final blow in the collapse of Yugoslavia was dealt by international rather than national decisions.[11] Where the IMF and the World Bank played a role in the dissolution of Yugoslavia was through refusal to relieve the inflation-fueled ballooning debts.

In the 1990s, the West supported non-socialist leaders, parties and governments in the former Yugoslav context. Nationalism was ignored, as the West assumed that newly formed governments would remove all associations with socialism and embrace the market economy. Moreover, and according to Susan Woodward,

> [b]y accepting the principle of national self-determination for the independence of states – without regard to the Yugoslav conditions of multinationality and the shared rights to national sovereignty of the Titoist system, or a willingness to enforce their unilateral decision on borders – Western powers were making war over territory inevitable.[12]

The wars turned into a cleansing of heterogeneity, and architecture was often used as a means of establishing ethnic purity.

Dislocation of Croatia

Croatia's balkanization from SFRY was bookmarked with the 87-day shelling and siege of Vukovar (August–November 1991) by the JNA, and the Serb forces' October 1991 bombings of the city of Dubrovnik, the first of these taking place on 23 October. That the bombings were globally reported in the mass media is not a surprise considering that Dubrovnik was recognized in 1979 by UNESCO as a World Heritage Site. This walled medieval city is populated with museums and libraries that map the founding of the city in the 7th Century, including its connections to Byzantium and the Franciscan and Dominican orders. Its notable buildings within the Old City are prime examples of Gothic-Renaissance palaces, Baroque churches and Franciscan and Jesuit convents. The destruction of such heritage by the JNA in Croatia prompted the use of the term urbicide by the architect and former mayor of Belgrade Bogdan Bogdanović in 1993 to describe the massacring of urbanity.[13] The 1991 use of urbicide had a different genesis to the earlier Cold War use of the term,[14] as it was related to war and the fragmentation of the ethnically heterogeneous SFRY. The urbicide of the SFRY and the destruction of its cities led to the first international involvement in the form of UN troops. The war-affected Yugoslavia had become a question of international security and one for the international media.

In respect to Dubrovnik alone, Bogdanović saw it as an attack "of a madman who throws acid in a beautiful woman's face and promises her a beautiful face in return."[15] Generally, he addressed the destruction in a way that distinguished between those dwellers who understand and revel in the heterogeneity of a city and those who are narrow-minded and hate the city; the mentality found in village dwellers or those who live on the outskirts of the city. He appeared to associate urban and anti-urban spaces, and those who dwell there, with particular sets of ideals and values including security, order and quality of life. At another level, he presented the "killing" of the city as the destruction of civic value, the elimination of the city's material foundations and commonalities, and the opportunities these afford for heterogeneous cultures to share public space.[16] Perhaps the differentiation was spurred by the 2 May incursion that occurred in the Serb village of Borovo Selo – located along the periphery of Vukovar – when a number of Croatian police officers and Serb civilians died. The deaths occurred in the wake of the arrest of two Croatian police officers by the Serb civilians in this village on the charge of discriminating against Serbs. The shortcoming of Bogdanović's suggestion is that he views the periphery and the peasant through the conservative lens of folklore whereby peasants are narrow-minded, backward and nationalistic, while city dwellers are educated, cosmopolitan and open-minded. The understanding of the violence perpetrated upon cities such as Vukovar and Dubrovnik facilitated a reading of the violence as a "revenge of the countryside" on the urban.[17] A backwardness supposedly found only in the rural culture undeniably evoked the Balkanist rhetoric of barbarity.

According to UNESCO, the intent of the targeting of Dubrovnik was psychological since projectiles directed towards the Old Town "were incapable of inflicting much actual structural damage though they did make an impressive and deafening metallic noise as they exploded and a small but sinister whistling sound as they flew over the roof-tops."[18] Moreover, the images used for media purposes were predominantly of burning yachts and boats, the smoke enhancing a scene of devastation and destruction.[19] Undeniably, watching yachts burn imparted a heightened psychological effect. Significant damage was suffered by the following religious and cultural buildings: St. John's Fort, the Synagogue, the Jesuit Church and the west windows of the Dominicans' Monastery and Museum.[20] The violence that was reported in the international media in the lead-up to the disintegration of SFRY and immediately thereafter contributed towards support for Croatia's independence and a general understanding of the destruction occurring in Croatia. The media portrayed violence perpetrated by the JNA and Serb forces often eliminated the violence directed towards the Serbs by the Croatians in

Dubrovnik and the rest of Croatia. The rising divisions between the ethnicities and republics were made possible through the rhetoric and threat of Other.

Croatia's desire to dislocate from the SFRY was voiced in a late 1980s speech given by the historian and retired general, and founder of the nationalist and right-wing Croatian Democratic Union (HDZ – *Hrvatska Demokratska Zajednica*), Franjo Tudjman, when he declared that "[t]he NDH [*Nezavisna Drzava Hrvatska* – Independent State of Croatia, as it was known during WWII when it was a Nazi state] was not simply a quisling creation and a fascist crime; it was an expression of the historical aspirations of the Croatian people."[21] The historical goal was possibly in reference to "long-standing Croatian aspirations for statehood – the 'Thousand Year Old Dream,' as it came to be known."[22] Following the election of Tudjman as the Croatian president in April 1990, a significant change was made to Croatia's constitution in December of that year. Croatia would no longer be legally bound to recognize the Serbs in Croatia, who formed approximately 12 per cent of the population, as constituent peoples; instead, they became a national minority.[23] This legal shift significantly affected the rights of the Serb population. Croatia declared independence and full sovereignty in June 1991.

The political changes in Croatia also coincided with incidents that broke out in May 1991 in Knin, a city with a majority Serb population and which between 1991 and 1995 acted as the capital of the newly formed Republic of Serbian Krajina (*Republika Srpska Krajina* – RSK). This Serb majority within Croatia predominantly lived along the borders with Bosnia-Herzegovina and Serbia. Their move to this area was made possible during the 16th Century; it was a military frontier separating the Habsburg and Ottoman empires. The Serbs were given the land by the Habsburgs, and the freedom to cultivate it, in exchange for defending the military border. Even after the collapse of the Ottoman Empire, the Serbs in this region retained relative independence; their desire for self-sufficiency and rural-oriented lifestyle also meant that the zone remained the poorest in Croatia. Their relative self-sufficiency was put into question in 1991. That year, protests broke out in Knin as a result of the reinstatement of the chessboard flag and coat of arms, the insignia used by the Independent State of Croatia during WWII. Even though the Germans set up the state, the Croatian Ante Pavelic led it. During this time, hundreds of thousands of Jews, Romani and Serbs were killed or interred in one of the 26 concentration camps in Croatia.The Serb media in Serbia seized on Croatia's reintroduction of perceived fascist symbolism to amplify the rhetoric that Croatia was attempting to repeat the violent acts perpetrated during WWII. From the late 1980s, systematic discrimination against its Serbian population included eradicating housing and employment rights, while the removal of Serb-oriented sociocultural factors such as ethnic language (the use of Cyrillic script), speech and literature was also noted.[24] According to a UN report:

> [d]ual script road signs even in areas where Serbs were a majority were torn down, Serbian sounding words were purged from the official no longer Serbo-Croatian language, some Serbs were asked to sign loyalty oaths to the government and some lost their employment in government positions or were subjected to confiscatory taxation.[25]

The period from 1991 to 1993 also saw 10,000 houses mainly owned and occupied by Serbs razed or blown up, sometimes while people were still inside.[26]

The violence occurring in Croatia in the late 1980s and early 1990s occurred at a time when the Serbian leader Slobodan Milosevic began his rise to power. Undeniably, he exploited the situation in Croatia to advance his power through rhetoric that insisted measures needed to be taken for the Serbs to feel safe in the Yugoslav republics. Milosevic secured victory in the 1990 presidential election in Serbia on the basis of a nationalist campaign spurred by violence between Serbs and Albanians in what was then known as the Autonomous Province of Kosovo and Metohija. Support for Milosevic increased after the speech he gave in Kosovo on 28 June 1989, on the 600th anniversary of the Battle of Kosovo at the Field of Blackbirds where the Turks defeated the Serbs and the medieval Serbia. The gathering was organized by the Orthodox Church. During the speech, he invoked the long history of the Serbs as a heroic and strong nation, even when under foreign rule. What Milosevic managed to achieve with this political move was to short-circuit time by bringing events of the distant past into the present, as well as use the mythologized narrative of the battle (the Serbian Prince Lazar dying whilst defending Serbia against the Turks), to deploy the event to proclaim heroism, strength and oppression as recurrent cycles in Serbian history. Similarly, by invoking the event and the name of Prince Lazar he placed the 1980s turmoil of Yugoslavia and his own persona along the same trajectory; not only was history repeating itself, but also Milosevic was the latest defender of Serbia in the present day.

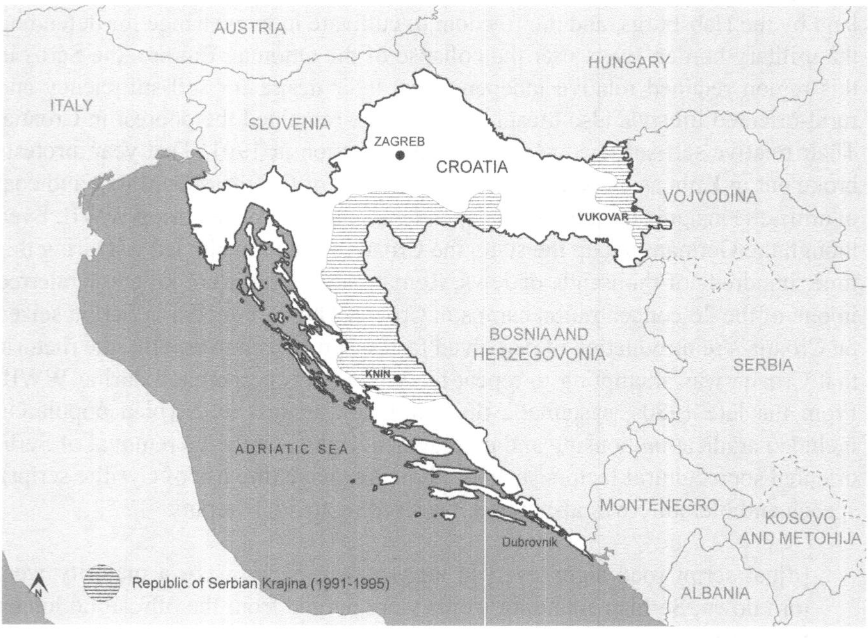

Figure 3.1 The boundaries of Croatia and the Republic of Serbian Krajina during the 1991–95 period.

The underlying agenda undeniably intended to revive nationalism by setting up Serbs as victims whilst also promoting a belief that Serbs may once again be forced to leave Kosovo, which historically was seen as their land.[27] The situation was exploited by Milosevic; he was quoted as saying "[y]ou shouldn't abandon your land just because it's difficult to live [sic. there], because you are pressured by injustice and degradation."[28] The speech, loaded as it was with emotive words, skated along the slippery line between fact and fiction. The rhetoric he used laid the ground for the forthcoming legal and constitutional changes which reached their apex with the eradication of the autonomy of Kosovo. Considering the significance of the Battle of Kosovo in Serbian history and mythology, and its association with the 500-year-long Ottoman rule, the speech incited initial support for Milosevic not only in Serbia, but also in areas in Croatia and Bosnia-Herzegovina with a significant Serb population. His address became a discursive and performative tool to fuel nationalism in collective Serb consciousness and to silence all those who were opposed to nationalism by manipulating the sociopolitical crisis beyond the territorial borders of Serbia; the aim was to amplify and solidify Milosevic's stronghold in Serbia and Serb-populated areas in Yugoslavia. The fact that Milosevic's message was both nationalist and socialist – not, as in Croatia, nationalist and market economy–oriented – gained both initial and ongoing support for him.

The war on homogeneity was not just a war of economics, international involvement and media manipulation. After all, the international community played a role by recognizing independence of Croatia. It was also a war over rights to land and territory. For Milosevic, the violence was driven by a seemingly 'given' necessity (and right) to join together Serb-populated areas in Croatia and Bosnia-Herzegovina with Serbia. For Tudjman, it was oriented towards the forced ejection of the once-constituent Serb nation from Croatia, which materialized in August 1995 as part of Homeland War (*Domovinski Rat*), a two-part offensive.[29] The first of these took place in May 1998, a part of Operation Flash (*Bljesak*), which restored control over the northern parts of the RSK, and the second was Operation Storm (*Oluja*), whose magnitude was more severe. The Operation Storm that forced over 200,000 Serbs to flee was presented as war for Croatian independence and as a question of taking back control of a significant chunk of Croatian territory measuring nearly 10,500 square kilometres, or about 18.5 per cent of the land in Croatia. The objective of homogeneity through balkanization, that is, imposition of 'ethnic purity' and assertions of rights to land that were associated with Tudjman's and Milosevic's rise to power, is both ironic and paradoxical considering the centuries-long cultural heterogeneity found in this region.

Dislocation of Bosnia and Herzegovina

Since Bosnia-Herzegovina was more multiethnic than Croatia, and prior to the disintegration of the SFRY considered a Yugoslavia in microcosm, efforts to eradicate heterogeneity there were more extreme. There were three groups in this war and three self-declared areas: Bosnian Muslims (Bosniaks) who were devoted to the

Republic of Bosnia and Herzegovina (October 1991); Croats who were faithful to the Croatian Republic of Herzeg-Bosnia (established in November 1991), and Croatia; and Serbs who were committed to the Republic of Srpska (proclaimed in January 1992) and Federal Republic of Yugoslavia. The Serb military action was undertaken by Bosnia and Herzegovina's Serbs, led by Radovan Karadzic and Ratko Mladic, with support from Serbia's Slobodan Milosevic. Military action on behalf of Croatians was led by Bosnia-Herzegovina's Croat Mate Boban and supported by Croatia's Franjo Tudjman. Muslim military combat was led by Bosnia-Herzegovina's Muslim first President Alija Izetbegovic. The heads of state and their parties had, just like in Croatia and Serbia, been elected on the basis of nationalist campaigns. Any opposition was met with processes of marginalization and silencing of the voices that deviated from the prevalent rhetoric of each ruling party. With the international support, Bosnia and Herzegovina held a referendum in February–March of 1992 for purposes of secession. The event led to unrest, the parcelization of Bosnia-Herzegovina and dislocation from SFRY. The Bosnian Serbs wanted to remain under their local leadership in association with Serbia, while Bosnian Croats wanted to become a part of Croatia. For the Bosnian Muslims, this was an opportunity to form a Bosnian Muslim independent state for the first time since their conversion to Islam during the Ottoman rule of Bosnia.

In Bosnia and Herzegovina, the destruction of cities would be widespread, but the cities whose destruction attracted most international attention were Mostar and Sarajevo. In terms of Mostar, the targeting of the historically significant 16th Century Old Bridge (*Stari Most*) over the Neretva river and nine other bridges inside the city meant that there was a real disruption to daily life for the locals.[30] The aim, according to Martin Coward, was to alter day-to-day routines and social interactions in such an extreme way that it became impossible for them to return to normal.[31] The Old Bridge was significant for its architecture and its narrative of a bridging of east with west. The symbolic connotation continued over time in that its targeting by the JNA and actual destruction by the Croatian Military Defence (*Hrvatska Vojna Odbrana* – HVO) in November 1992 symbolized the breakdown of communication within multiethnic Bosnia.

The flattening of heterogeneity was also present within the built environment, which was adorned with Ottoman, Mediterranean and Western European architectural elements. The documentation of urban destruction was recorded and commented upon by a group of architects in a document called *Mostar '92 – Urbicide*.[32] However, the targeting was not just directed towards destroying public buildings, community housing, urban infrastructure, Catholic churches, Austro-Hungarian monuments and public parks and gardens. It was also to erase the place of rest and memory of those who passed away, whilst simultaneously using the city as a site of death in the name of eliminating diversity. The architect Krešimir Šego wrote that in Mostar "[t]he cemeteries were destroyed with the town, and the dead are now buried in the parks and gardens."[33] Zoran Bošnjak, another architect, was more specific in describing the elimination of diversity once present in this city. He noted that "[t]he little park of Liska, where sweethearts used to meet in the shade of age-old pines, is now a crowded graveyard of war victims – young and old, military and civilians, men and women, Serbs, Croats and Muslims."[34]

Sarajevo, as the capital of Bosnia and Herzegovina, also received considerable media attention. The focus was not necessarily due to its status as capital, but more for what it was seen to signify – a melting pot of cultures and ethnicities. Its siege by the Serb forces was thus perceived as the flattening of an open-minded and plural Europe, and its destruction described as "warchitecture" by Sarajevo's Association of Architects.[35] In 1993, in the midst of the siege of Sarajevo, this destruction was marked with an exhibition in Paris *Warchitecture – Urbicide Sarajevo*. Carefully detailed maps indicated damage inflicted on specific architectural elements and spaces as well as noting whether the destruction was partial or complete. The exhibition displayed architecture as a spatial and visual manifestation of damage by the Bosnian Serb army where supplied maps and photographs framed architecture in terms of war ruins. Targeted architecture was also broken down into historical and cultural categories: Ottoman, Austro-Hungarian, modernist and contemporary.[36] The project had its international launch at the Paris Pompidou Centre in 1994; the frontier of war was expanded virtually through the exhibition, including the borders of Sarajevo as a city under siege.

A well-documented example of "warchitecture" is the destruction of Vijecnica; the 19th Century Sarajevo City Hall / the National and University Library of Bosnia and Herzegovina. The 1896 building is located in proximity to Bascarsija – a historic bazaar built in the 15th Century during Ottoman rule – and the residential zone on the other side may be seen as gateway structure between these two zones. The imposing scale of the structure underwent a couple of redesigns; the cultural diversity was present with the choice of the architects. The first design was by the Czech architect Karel Parik. However due to disagreements, presumably on matters to do with finance, Parik was replaced with the Austro-Hungarian architect Alexander Wittek. His role was to rework Parik's proposal. The pseudo-Moorish-style design was inspired by the style of Islamic mosques; the inspiration was an interpretation and one tailored to suit the aesthetic of the Austro-Hungarian Empire. The completion was seen through by the Croat architect and conservationist Ciril Metod Ivekovic following Wittek's death in 1894. Since its completion, the structure stood paramount in the identity of Sarajevo.

In August 1992, Vijecnica was attacked over a period of three days; Bosnian Serb grenades carved up the structure. The targeting of the 19th Century pseudo-Moorish-style structure also destroyed centuries-old records from the Balkans, over a million books and 100,000 manuscripts. It is said that

> [a]ll over the city, sheets of burning paper, fragile pages of grey ashes, floated down like a dirty black snow. Catching a page, you could feel its heat, and for a moment read a fragment of text in a strange kind of black and grey negative, until, as the heat dissipated, the page melted to dust in your hand.[37]

The destruction was associated with destruction of memory – memoricide – as it was an attempt to destroy the documentation of history.[38] In other words, the shelling of significant buildings such as Vijecnica or the Oriental Institute contributed to the historical loss of numerous documents in Arabic, Persian and Hebrew, and symbolically to an eradication of coexistence. However, this was an attempt to eliminate not only history but also daily habits, as seen in the targeting of everyday places such as markets and cafés.[39]

The destruction of buildings in Eastern Sarajevo (largely populated by Bosnian Serbs), however, largely escaped global media attention and interest. To this day, the city is divided into two parts: Western and Eastern Sarajevo. Balkanization in this instance was deployed for purposes of destruction, division and ultimately reconstruction; motivated by the creation of ethnically pure states and histories. From this perspective, the infliction of balkanization cannot be disentangled from questions of how to deal with the rubble of a city and reconceive a new identity by integrating both the distant and immediate history and memory of a place.

Stability and peace

Remaking as falsifying history

The remaking of cities in the aftermath of war was not a straightforward scenario of reconstruction for the sake of allowing people and life to return back to 'normal.' The reconstruction was tied to a number of political and 'peace'-driven initiatives that posited peace as contingent on creating ethnically homogeneous enclaves; the process was highly contrary to the historical heterogeneity found in this region. In other words, the aftermath was managed through processes of administration, and where history became a significant tool in pushing through this agenda whereby falsification of history has been congruent with the denial that there is more than one story, one history.

The collapse of RSK in Croatia in 1995 led to the Erdut Agreement for the region of Eastern Slavonia, in which the city of Vukovar is also located. The Agreement facilitated 'peaceful' reintegration into the Republic of Croatia, a process that lasted from 1995 to 1998. In practice, stability was generally achieved by the Croatian Army driving out the majority of its Serb population during the 1995 military operations. The overall military strategy and specific techniques of this operation are not only studied within the US military for training purposes, but also portrayed as a desirable approach to politically altering history.[40]

Because Bosnia and Herzegovina did not have one clear ethnic majority, the task of creating an ethnically homogeneous geopolitical entity there was significantly harder. The outcome was the creation of two sub-states managed by the UN protectorate as a third party: the Federation of Bosnia-Herzegovina (FB-H), predominantly Bosnian Muslim and Bosnian Croat and constituting 51 per cent of the territory; and the Serb Republic (*Republika Srpska*), occupying the remainder 49 per cent and predominantly Bosnian Serb. The international 1995 Dayton Agreement perpetuated balkanization through an establishment of homogeneous enclaves, whilst simultaneously creating the conditions for Bosnia and

Herzegovina to become dependent for its survival on the international community (NATO, the US, the World Bank, IMF and the EU).[41]

Whilst the signing of the Dayton Agreements at Versailles in 1995 is regarded as synonymous with peace, pluralism and reconciliation, the reality seems to be very different. The lack of ethnic plurality and cultural diversity is reinforced by the fact that the agreement was contingent on parcelizing the FB-H into ten cantons, with little mixing between different ethnic groups. In other words, from 1995, about 90 per cent of its people have lived in homogeneous communities. The return of displaced peoples to areas from which they were expunged – the Serb Republic and the FB-H – has been nominal.[42] For a country that was multi-ethnic for centuries, forced balkanization is an instatement of violence, although less visible than the one seen during 1992–95 which, apart from notable destruction of the built environment, also saw the displacement of 1.5 million people and over 100,000 deaths in this former Yugoslav Republic.

Current violence is steeped in deception through rhetoric that invokes democracy and peace, and obfuscates the barbarity and violence that is synonymous with this Balkanist zone. The belief that the one is definitely replacing the other through the bestowal of economic loans along with associated administrative policies and legal conditions is a means to veil the political impositions of control; to believe in such a binary rhetoric is the greatest

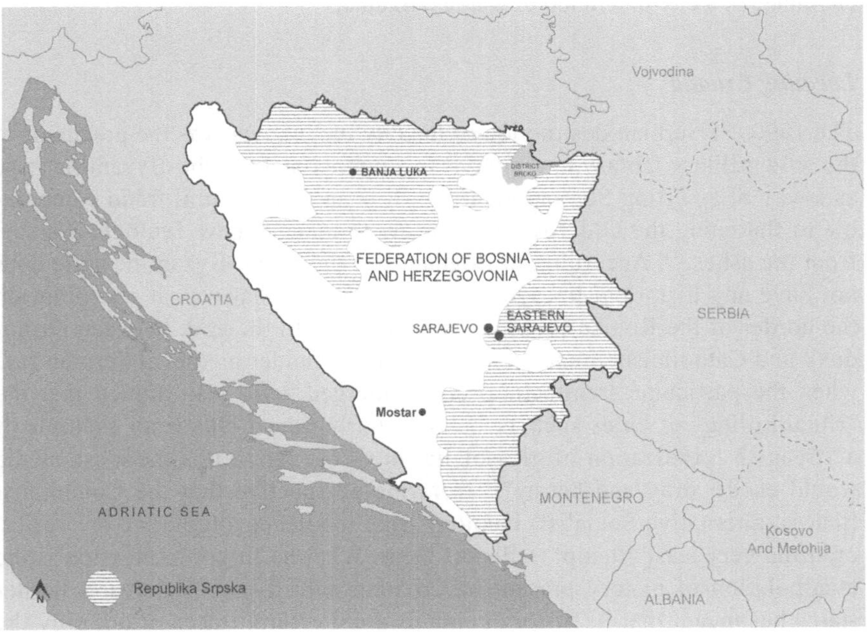

Figure 3.2 Present boundaries of the Federation of Bosnia-Herzegovina and Republika Srpska (The Serb Republic). Brcko District was formed in 1999 as a self-governing unit to reflect the multiple ethnicities present in the city of Brcko and the surrounding areas even after the war.

moral narcotic. While this region has been under many rulers, including those of the Ottoman and the Austro-Hungarian empires, current violence and control is more dangerous precisely because of the rhetoric of democratic and plural solutions brought upon this region. Integration of the Balkans, that is, Western Balkans, with Europe and the West is made contingent on reducing diversity to circumvent the supposed difficulties that arise when negotiating multiple identities and values in one territory. Identity formation through the implementation of balkanization is thus equated with reactivating and boosting the necessity for territorial enclaves of ethnic homogeneity, rather than diversity. What this thinking simultaneously shows is a belief in the superiority of Western spatial formation on the basis of ethnic-separatist thinking and its own short-sightedness in the face of historical demonstration that ethnic diversity and coexistence have been possible in this region.

At a national level, balkanization makes possible the use of specific historical events to craft memories of recent wars so as to lock the present within a particular framing of the past. The reconstruction of Croatia and FB-H is a complex cross-fertilization of myth, fabrication and selective amnesia. The new identity is one of truth-formation created for purposes of dwelling upon the recent past; the political dogma insistent on *not forgetting* is there to keep the recent memories and emotions of the war raw and alive. The symbolism of war is a powerful tool in not only mobilizing history for purposes of achieving ethnic homogeneity and for keeping the nationalist flame alive, but also in facilitating a perpetually despondent image of a city in a limbo of war and trauma.

Locating Croatia

The 1990–91 Serbian destruction of the city of Vukovar was the precursor to the wild military fantasy to rebuild it in a non-existent Serbo-Byzantine style as a way of asserting Serbian identity.[43] It is an architectural fraud comparable to flattening the Old Town of Warsaw to raise "a new Teutonic Warsaw from the ashes."[44] According to Bogdanović, the Serbo-Byzantine style does not have any historical lineage in Vukovar; a city that was at one point on the border of the Roman Empire – associated with the split between Orthodoxy and Catholicism – and one that was never under medieval Serbian rule when the particular architectural style emerged.[45] He also argued that the "ritual killing" of cities such as Vukovar during the 1990s is an example of a "freakish 'civilization of glossolalia' in whose labyrinths the mass media would easily stray and betray"[46] from the perspective that the Communist (that is, alternative Socialist) ideology tried to silence alternative memories, or at the very least attempt to distort them. What he forgot to mention is the manipulation of history present in Croatian narratives.[47] Vukovar's border status has meant that its history contains architectural traces of not only the Austro-Hungarian Empire, but also the Independent State of Croatia, Yugoslavia (during the Kingdom and the Socialist period), the RSK and now the

Republic of Croatia.[48] Oddly, reconstruction post-1995 has not only strengthened balkanized ethnic enclaves but is also consistent with removing any trace of history that does not complement the current ideology of the Croatian Republic.

Using Bogdanović's own logic of "ritual killing" of cities as motivated by a distorted image of events, memory and history, an alternative narration of recent destruction and reconstruction of Croatia is also possible. This alternative understands that the violence within the SFRY context is complex and that a clear perpetrator-victim identification is not possible, just as the destruction of architecture and cities cannot be separated from various political changes taking place at that time. More broadly, the naming of a Croatian "Homeland War" is peculiar since it only ever existed as a nationalist project rather than as a territorial entity.[49] National identity is managed by bringing together nationalism, sovereignty and ideology for purposes of eliminating other ethnic narratives. That the battle of Vukovar is coeval with the birth of a Croatian nation is presented in a wide variety of media; from music to education, ceremonies, currency, street names, stamps and religious heritage.[50]

Figure 3.3 Prominent war memorials in Vukovar: 1 – War Museum inside the Vukovar Hospital; 2 – Water Tower; 3 – Monument to Victims of Homeland War; 4 – Memorial Cemetery for the victims of Homeland War; 5 – Ovcara Memorial Centre.

Vukovar is the example par excellence of memorializing a singular perspective of the past. This is done through monuments, with Vukovar and the region of Slavonia seen to count the most: the Ovcara Memorial Centre, Monument to Victims of Homeland War, War Museum-Vukovar Hospital, Memorial Cemetery for the victims of Homeland War and the Water Tower being the most prominent and promoted.[51] The Ovcara Memorial Centre is used by Croatians to commemorate the

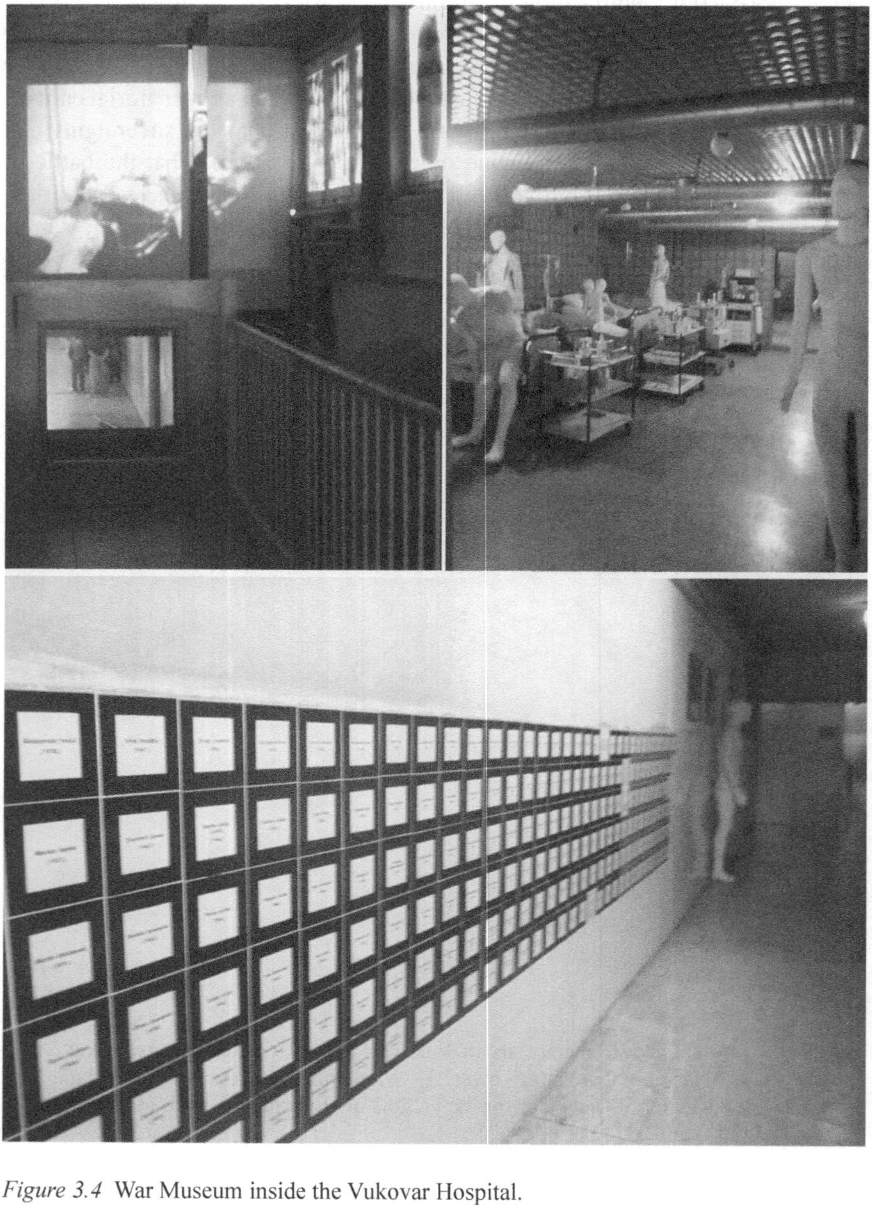

Figure 3.4 War Museum inside the Vukovar Hospital.

start of the siege when at least 200 Croatians were massacred near Ovcara by the JNA and the Serb paramilitaries. The relatively closely located Memorial Cemetery for Croatian defenders commemorates the Homeland War whilst ignoring the ethnic cleansing of Serbs in 1995 and massacres perpetrated since 1991. There is a meta-rhetoric that history can be understood only in black-and-white terms, the position that is used for purposes of establishing truth, that is, historical amnesia.

The discourse of truth is the capacity to strategically manoeuvre a battlefield of narratives when there is only just enough fact to validate all the erasures. Simultaneously, truth is contingent on the dominant Western and Anglo-American privileging of violence perpetrated during the dissolution of Yugoslavia. What makes this partisanship even more peculiar is that the recommendations made to member governments of the Council of Europe on the teaching of history in the 21st Century include recognizing it as a vital element "in the promotion of fundamental values, such as tolerance, mutual understanding, human rights and democracy."[52] It is unlikely that Croatia as an EU member abides by these values, or at least does only enough to send up a smokescreen. Ethnic plurality is seemingly accepted in the city of Vukovar, given that this is one of the few areas in Croatia with a small Serb ethnic minority. In meeting the conditions of the European Charter for Regional or Minority Languages, bilingual plaques are displayed there in both Latin (Croatian) and Cyrillic (Serbian) scripts. However, divisions are present in most other areas of life. Coexistence is contingent on primary schools having two entrances, one for the Croatians and the other for the

Figure 3.5 Memorial Cemetery for the victims of Homeland War.

Figure 3.6 The unreconstructed Water Tower as a signifier of the war.

Serbs. Classrooms are also separate, as are the ways in which history is taught. If anything, such an approach teaches segregation and an inability to see other people beyond ethnicity. There is a fine line between addressing violence and history in order to understand the history of violence, using history to spearhead acceptance of the other or using it to bear hatred. Benjamin's thinking resonates in this context: suffering and violence from the past is often used to enable hatred rather than reconciliation.[53]

Memorializing history is dependent upon falsifying historical moments. This is seen in the memorial erected by Croatians in the majority Serb village of Borovo Selo to commemorate the death of 12 police officers who died in a violent encounter with Serbs in 1990, an incident that has been presented as the precursor of the violence in Croatia. The memorial was vandalized by the Serbs, probably because they were in disagreement with its one-sided interpretation of history. The Croatian response was the erection of a new memorial as a way to have the final say on Croatian power and territorial occupation.[54] While the significance of war memorials and war heritage is not a new concept, with memorials and monuments deployed to construct historical framework and identity from Poland and France to Cambodia,[55] the specificity of Vukovar is that war has been subsumed into tourism in a way that undermines the possibility of facilitating other forms of tourism, from archaeological settlements of Troy to the nautical culture.[56]

War and trauma occupy the political and built landscape, implemented not only through segregation and the supposed inability to coexist with ethnic minorities but also on behalf of the military defenders of Vukovar. Nationalism is kept alive by memorials coded with simplistic narratives and symbols from the war. That war symbols and names extend beyond the boundaries of Vukovar is seen with the 2017 construction of its airport terminal and the renaming of the airport in Zagreb to Franjo Tudjman Airport, a name suggesting that Croatia meets the world in a celebration of war criminals and nationalist tendencies. The design of the terminal building, which arose out of the competition organized by the City of Zagreb in 2008 and won by Branko Kincl, Velimir Neidhardt and Jure Radic, celebrates the symbolic national insignia even in landscaping, with a chessboard landscaped garden plot fronting the

Figure 3.7 Ovcara Memorial Centre.

Figure 3.8 Monument to the Victims of Homeland War.

main entry. The significance of Franjo Tudjman is also seen with the state initiated and funded 2018 monument erection to this former President in the city centre of Zagreb. Unveiled on the 19th anniversary of his death, and amidst the presence of the current Croatian President Kolinda Grabar-Kitarovic who also gave a speech which sided Tudjman with greatness and honour, this more than four metre high statue is a symbolic signifier that extreme acts of human rights violations are now associated with heroism and democracy.[57]

The destruction that Dubrovnik incurred has been remediated and is largely not visible. However, upon entering the Old Town, one encounters large boards affixed to the walls that encircle it and note the complete or partial destruction incurred during the 1991 bombardment of Dubrovnik by the JNA; for the most part, these indicate damage to areas of pavement and rooftop strikes rather than actual destruction. One of the prominent buildings destroyed was the Baroque-style house that was home to the painter Ivo Grbic. The reconstructed building has now been turned into an exhibition space for the destruction incurred. The Memorial Room of the Defenders of Dubrovnik is another notable reference to the war, a space found on the ground floor of the Gothic and Renaissance style Sponza Palace built in the 16th Century, to commemorate all the lives lost in the defence of this city between 1 October 1991 and 31 May 1992. Its interior is populated with photos, live video footage and memorabilia in the form of war objects and a burned flag.

A more charged evocation of the war is found in Dubrovnik's Museum of the Croatian War of Independence. The museum is located within the walls of the fort on Mount Srdj. The space is broken into four areas, each of which commemorates a

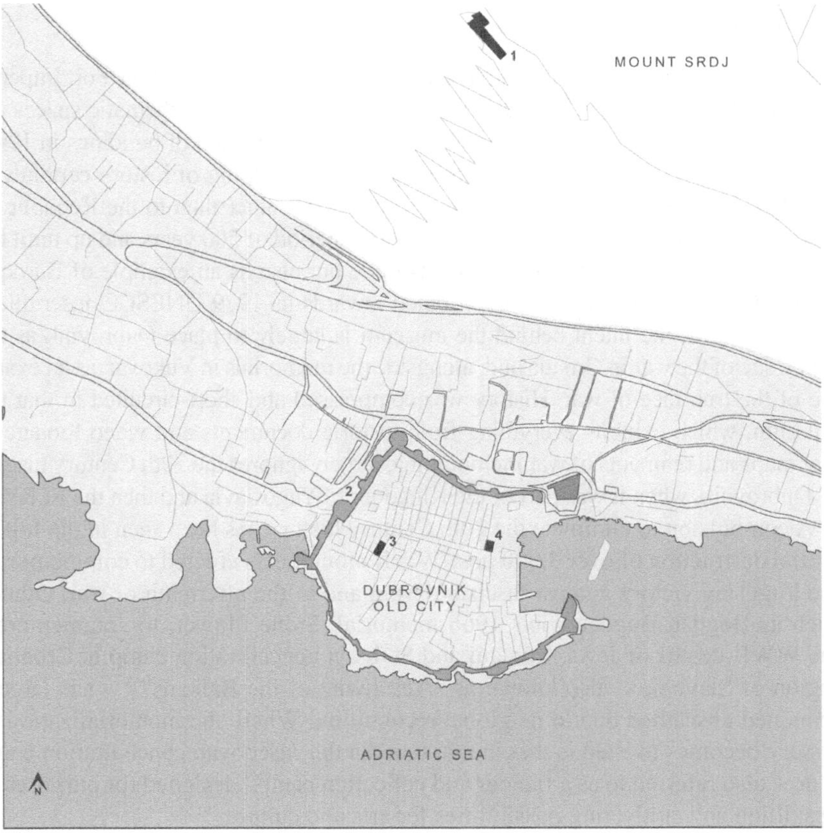

Figure 3.9 Prominent memorials and reminders of the war in Dubrovnik: 1 – Museum of the Croatian War of Independence; 2 – Old City Walls – large boards on Old City Walls note the destruction incurred during the 1991 bombardment; 3 – Ivo Grbic Art Gallery; 4 – The Memorial Room of the Defenders of Dubrovnik.

Figure 3.10 Façade of the Sponza Palace in which the Memorial Room of the Defenders of Dubrovnik is located, and inside the Memorial Room.

different aspect of history. The first maps the 19th Century history of the Fort Imperial and the fall of the Republic of Dubrovnik, while the other three thematic spaces are oriented towards commemorating the targeting and destruction of buildings in 1991, civilian suffering and the liberation of the city. Its weighting of history certainly is directed towards the 1990s period and its aftermath, rather than to the Republic of Dubrovnik (then known as Ragusa), which for a period of 500 years and up until the 19th Century was known not only for its trade but also as an example of Baroque urban planning, features which contributed towards its 1979 UNESCO inscription. In other words, the intent behind the museum is largely to place Dubrovnik as the prime site of the war in Croatia and, alongside the memorials in Vukovar, as an example of the true face of war. History was compressed and short-circuited in that the museum, which exhibits everything from historic documents and video footage to war maps and remnants of war memorabilia, largely ignores the 20th Century history of Dubrovnik, when it was part of the Kingdom of Yugoslavia and then the SFRY.

A compulsion to eliminate the 20th Century history has been seen in the full or partial destruction of over 3,000 post-WWII monuments erected to commemorate the Yugoslav victory against fascism during and in the aftermath of war. Others, such as Bogdan Bogdanovic's 1966 memorial, Stone Flower, to commemorate the WWII deaths of Jews, Romani and Serbs in concentration camp in Croatia's region of Slavonia – also known as "Auschwitz of the Balkans"[58] – has largely remained unscathed due to its global recognition. Where the memorialization of history becomes twisted is that in mass media the Jasenovac concentration camp is now also referred to as a 'labour and collection camp,' designed for purposes of upskilling and cultivating possibilities for arts and culture.[59]

Even the Memorial Museum Jasenovac, designed by Helena Paver Njiric and added in 2006–07 for purposes of supposedly addressing the fascist past, denies the attempt to collectively remove particular national identities.[60] It seems that the dominant act of destroying anti-fascism monuments in Croatia is simultaneously synonymous with a tacit agreement with fascism, as well its denial. Presumably, the EU recognized the significance of this (continuing) intent and operation at the time it welcomed Croatia into the European fold. Memory not only has been militarized and state-initiated-funded but also delegitimized and short-circuited. When it comes to the SFRY period, memory and history can be recalled and addressed only from the perspective of anti-communism and anti-Balkanism. That the need to remove any association with the Balkans and Socialism was paramount was seen in the 1990s name change of the movie theatre in Zagreb from Cinema Balkan (*Kino Balkan*) to Europe (*Europa*).[61]

Locating Bosnia and Herzegovina

The transition to peace in Bosnia-Herzegovina has been related directly to urban post-socialist reconstruction, largely made possible through foreign funding. As for Mostar, reconstruction remains tainted with divisions and disagreements between the Bosnian Croats, who reside in the western part, and Bosnian Muslims, who inhabit the eastern part. This is despite the reconstruction of the Old Bridge in the old Muslim quarter, held up as an icon of reconciliation between the two warring sides. The impetus for the reconstruction of the bridge came from UNESCO in 1994; the funding flowed from numerous organizations such as the World Bank, EU, the Aga Khan Trust for Culture and the World Monuments Fund as well as in the form of donations from European countries such as France, Italy, Turkey, Croatia and the Netherlands. The funds were predominantly directed towards visible and symbolic projects such as monuments. Symbolically, reconciliation was directly representative of the opportunity to present glossy images to the world, rather than investing in projects that would alleviate the real war trauma within the region. The ongoing presence of trauma and divisions is evident in the stone plaque set below the bridge, reading "[e]xtremists HVO [*Hrvatska Vojna Odbrana* – Croatian Military Defence] and HV [*Hrvatska Vojska* – Croatian Army] destroyed this 427-year-old bridge."[62] And so that the international community need not be aware of it, or may avert its eyes from it, the plaque cannot be seen from the tourist trail.[63]

The borders in Mostar are not as obvious as those in the West Bank or Nicosia, but they are certainly present. Within this context, balkanization is found within the Old Bridge itself. Despite serving as a spectacular emblem of peace and coexistence, the actual bridge now reinforces the separateness of the two sides of the city. As it affects Mostar, the bridge restoration is no longer just about the

Figure 3.11 Prominent past and present memorials in Mostar: 1 – Bogdan Bogdanovic's 1965 Partisan Memorial Cemetery; 2 – Park Zrinjevac: the location of the Bruce Lee sculpture since 2013; 3 – Spanish Square: location of the 2005 Bruce Lee sculpture; 4 – Stari Most (Old Bridge) / Novi Stari Most (New Old Bridge).

symbolism of reconnecting the Bosniaks (Bosnian Muslims) with the Bosnian Croats. It is the way its restoration to 'original heritage' state is nothing more than a fabrication of history. Not only can the bridge not serve the same function as it did prior to its filleting, but also the craft does not match the originary tectonics, despite attempts at heritage accuracy by referring to numerous documents and sourcing stone from the same quarry the builders of the original used. The process of reconstruction even involved archaeological and scientific research that uncovered bridge structures predating the 1557 construction of the Old Bridge.

The actual construction, which began in 2001, was accompanied by a rhetoric of 'honesty and integrity' in ensuring the application of ancient techniques and methods. This is in light of the fact that every bridge that has existed in this location up until the 2004 unveiling of the New Old Bridge was constructed in a different style, without replicating the previous design. Tradition may need to be derived from a basis other than mere replication of the structure that was there prior to destruction. Addressing the typology of the bridge, its uses and weathering are more suited to framing the context of history than a reconstruction that treats history as a frozen object. The only difference that has been permitted in the rebuilt bridge is the name: once the Old Bridge, but now a New Old Bridge.

While Todorova has described Balkanism in terms of stages of development between developed and less developed areas, and a bridging together of cultures

Figure 3.12 The UNESCO-protected Stari Most (Old Bridge), that is, Novi Stari Most (New Old Bridge) in Mostar.

and spaces, a more precise description of the current state of FB-H demands greater complexity. Perhaps that state is best encapsulated by Georg Simmel's observation that "practically as well as logically, it would be meaningless to connect that which was not separated and indeed that which also remains separated in some sense."[64] Although the bridge is physically connecting the two parts, the politics of balkanization has it reinforce the inability of the two parts to coexist. But the fact that this is not just a question of the present operating as two separate islands in the face of the immediate past can be seen in the attempts made to remove any history when coexistence was possible. The identity of post-Dayton Mostar is dependent upon the dismantling of socialist and anti-fascist history, as seen through the damaging of Bogdanovic's 1965 Partisan Memorial Cemetery in 1992, which was built with the intent to commemorate the deaths of those who fought against fascism, in Mostar and across Europe and the world. The memorial has been only partially restored.

The attempts to breach divisions were also seen in the initiative by Mostar Urban Movement to construct a life-sized bronze statue of Bruce Lee. The choice of the martial artist was so far removed from the context of the war that it seemed an appropriate symbol of reconciliation and solidarity; Lee was a figure loved by all who grew up in SFRY. The sculpture, made by the Croatian sculptor Ivan Fijolic, was unveiled in November 2005 in the Spanish Square, only to be vandalized soon after. The memorial was removed and relocated to the nearby Zrinjevac park in 2013. Its placement

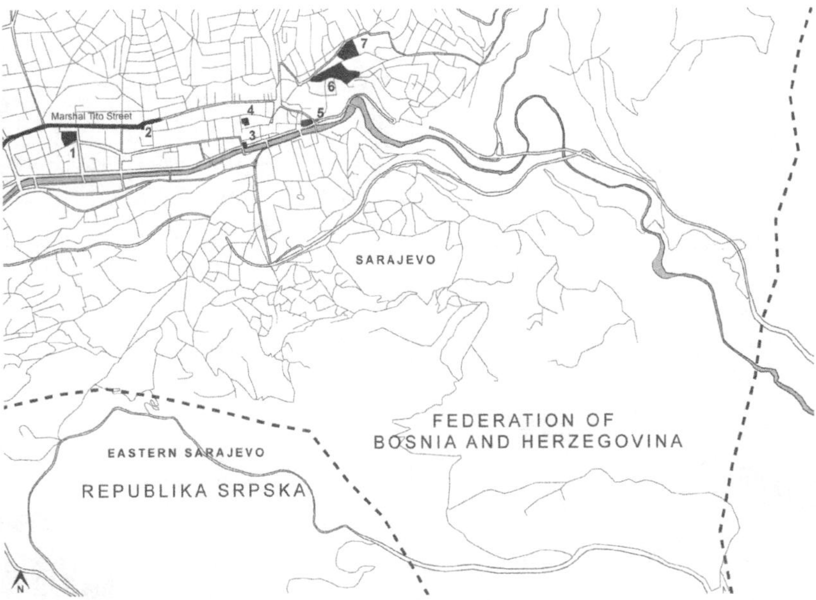

Figure 3.13 Prominent past and present memorials, and rebranded and reconstructed buildings in Sarajevo: 1 – Sarajka: BBI Centar; 2 – Eternal Flame memorial (Vjecna Vatra); 3 – Memorial to mark the assassination of Archduke Franz Ferdinand of Austria and Museum Building in 1914; 4 – Gazi Husrev mosque; 5 – Vijecnica; 6 – Martyrs Memorial Cemetery Kovaci; 7 – Museum of Alija Izetbegovic.

is significant, as the statue faces neither the western nor eastern side; in other words, Bruce Lee is not siding with either the Bosnian Croats or the Bosniaks.

When it comes to Sarajevo, the lack of commonality between the different ethnicities is exacerbated through reconstructed architecture. The reconstruction of Vijecnica (Sarajevo City Hall / the National and University Library of Bosnia-Herzegovina), and its symbolic association with the different ethnicities that have passed through, occupied and lived in the city of Sarajevo, have been eclipsed by the fact that attempts were not made to deal with the history of trauma; instead, the focus has been symbolic and power-oriented. The reconstruction was seen as a tribute to Europeanization and civilization, as values opposed to barbarism. This is significant considering that the building has been associated with the assassination of Archduke Franz Ferdinand in 1914 (some of the last visual records of his life were captured on the steps of Vijecnica) and indirectly with the start of the Balkans War, the outbreak of WWI and the association of the Balkans with violence and despotism.

The reconstruction and reopening of the building in 2014 seem to have erased much of its public use, considering that from 1947, Vijecnica was turned into a library, public space and a cultural seat of multiethnic Bosnia and Herzegovina. The reopened building has reverted to its 'original' use under a change of name to Sarajevo City Hall. However, this city hall is used largely for private events, with some spaces activated for token exhibitions to do with the 1990s war – largely for purposes of attracting tourists. In the attempts to recreate the 'original' structure and style, including interior patterns and colours – despite its 19th Century style being moulded to suit the

Figure 3.14 The reconstructed 19th Century Vijecnica.

Austro-Hungarian aesthetic palette – the chosen exterior and interior colour hues have been made so vivid that they seem plastic. The intent behind the chosen colours has been for each hue to retain its intensity, despite the inevitable loss of opacity that original colours undergo over time. The reconstructed building denies the wear of time.

The reconstruction also rests upon the widely held principle of ethnonationalism; at the very entrance of Vijecnica, a plaque in Bosnian and English has been placed reading,

Figure 3.15 The entry to Vijecnica – the plaque is a reminder of war and those seen responsible for it.

Figure 3.16 Inside Vijecnica – the exhibition as a reminder of war and destruction.

> On this place, Serbian criminals in the night of 25th-26th August 1992 set on fire the National and University's Library of Bosnia and Herzegovina. Over 2 millions of books, periodicals and documents vanished in the flame. Do not forget. Remember and warn![65]

The plaque is not only overturning the heterogeneity and plurality of views and values associated with Vijecnica prior to its destruction but also signposting that the building of the post-socialist identity of the capital is premised on cementing the polarization of public and private space as well as the perception of 'us as victims' and 'them as perpetrators.' The role of architecture in Bosnia and Herzegovina seems to be oriented not only towards erasure and selective narration of history but also colonization considering the association of Vijecnica with the Hapsburg era in an attempt to throw out the Balkanist past.

However, the reconstruction of Sarajevo is more complex than this, and any attempt to ascribe to it a particular identity may not be possible. The reconstruction of the city's Gazi Husrev Mosque was made possible by the Saudi government. The process of remaking was made provisional on eliminating any versions of Islamic design considered by the Saudi government to lack merit. Thus, the 1996 reconstruction of Gazi Husrev Mosque involved stripping the interior of all Balkan Islamic design and colour. Seemingly, to locate the historically multiethnic and multireligious Sarajevo is contingent on removing any sense of cultural context and

Figure 3.17 View from the Museum of Alija Izetbegovic towards the Martyrs' Memorial Cemetery Kovaci; the cemetery houses the bodies of the Bosniak soldiers killed during the 1990s war, and the body of Alija Izetbegovic.

difference in practicing and representing Islam. This is not a question of reinterpretation of Islamic design, a tactic implemented during the Austro-Hungarian reign when the Austrians reinterpreted Orientalism to create Vijecnica. That may well deserve to be classified as a sign of colonialism; but what is different in Sarajevo's Gazi Husrev Mosque is that reinterpretation is permissible only if it replicates a singular vision of Islam – the vision of the Saudi government. The funding corporations, and their ability to determine what needs to be preserved, drive the decisions behind not only reconstructions but also historical narrative. The guided tours within and around the historic Bascarsija bazaar now narrate the centuries-old buildings such as the Gazi Husrev Mosque in relation to the 1990s war rather than to its more distant history.

The national politics of FB-H are just as single-minded. Despite the Museum of Alija Izetbegovic being portrayed as a museum of plurality and inclusivity, the reality is significantly different. The agenda is geared towards setting up a clear victim – perpetrator mindset, where the perpetrator is the genocidal Serb whose violence is nothing short of Nazism.[66] Izetbegovic comes across as some kind of visionary and a preacher of tolerance, though a totalitarian one. Within the corridor that connects two parts of the museum, photos of Izetbegovic are interspersed with his sayings.

Alongside his prisoner mug shot taken in March 1946, during his imprisonment post-WWII, he is noted as saying "to openly look the truth in the eye is to realize that the meaning of life lies in the fight against evil."[67] The quote is directed against the ideology of the socialist rule in Yugoslavia post-WWII, as well as an

Figure 3.18 Inside the Museum of Alija Izetbegovic; the corridor features plaques with Izetbegovic's photos and words of wisdom.

attempt to establish his beliefs as the only voice of truth. Other photos and quotes permanently exhibited in the corridor largely depict Bosnia and its people as innocent in the face of the other side, represented by Milosevic and Tudjman. Izetbegovic is not only the face and voice of verity, but also to engage with and support him has the character of revelation and miracle, since, after all, "[s]weet lies are of no use, while bitter truths can have a healing power."[68] That the museum is in the immediate vicinity to the Martyrs' Memorial Cemetery Kovaci where a notable number of Bosniak soldiers who were killed whilst defending Sarajevo were buried, and where the body of Izetbegovic was also laid to rest, makes it questionable as to how these reminders of death and war can have a healing power.

Sarajevo largely seems to be constructing its identity through the 1990s violence, and in response to a need to remove the historical socialist tag. The post-socialist transition is intent on the removal of any communist names from the public domain. The socialist store Sarajka has also been reconstructed with Saudi funding. The glitzy new shopping centre not only has a new name – BBI Centar – but also articulates the instatement of religious cultural practices previously not followed by the Bosnian Muslims: there is a ban on serving alcohol inside. Another key target in the political manoeuvre to erase any trace of socialism has been street names, with 40 per cent of streets being assigned a new name.[69] One name change that did not go ahead, as a result of public protests, is to Marshal Tito Street. This is significant since that name was to be replaced with the name of former President Alija Izetbegovic. While such resistance may be

Figure 3.19 The rebranded Sarajka, now BBI Centar, on Marshal Tito Street.

seen as an example of nostalgia, the more radical agenda is found in the thinking that alternatives found in Titoist socialism have not been completely eradicated from the people's psyche.

Traces of more distant histories are still present, though few and far between. One of these is the Eternal Flame (*Vjecna Vatra*) memorial built in 1946 to commemorate the anniversary of Sarajevo's liberation from Nazi Germany and the fascist Independent State of Croatia. The flames have stood the test of time and are still burning despite the change of government and political orientations. The inscription at the 1946 Eternal Flame notes:

> With Courage and the Jointly Spilled
> Blood of the Fighters of the Bosnian-Herzegovinian,
> Croatian, Montenegrin and Serbian Brigades
> of the Glorious Yugoslav National Army;
> with the Joint Efforts and Sacrifices of Sarajevo's Patriots
> Serbs, Muslims and Croats on the 6th of April 1945
> Sarajevo, the Capital City of the People's Republic
> of Bosnia and Herzegovina was liberated.
> Eternal Glory and Gratitude to the Fallen Heroes
> of the liberation of Sarajevo and our Homeland,
> On the First Anniversary of its Liberation -
> a Grateful Sarajevo

Figure 3.20 The 1946 Eternal Flame memorial.

With the memorial located in the city centre, and designed by the recognized architect, urban planner and educator Juraj Neidhardt, its meaning sits uncomfortably against the plaques in the immediate city centre vicinity that identify the Serb destruction as criminal. One of these recent plaques is located a couple of hundred metres from the Eternal Flame memorial, located at the entry of a clothing store, Parfois. It was allegedly placed there by the citizens of Sarajevo and reads, "[a]t this place on 27 May 1992, the Serbian aggressors killed 26 Sarajevo civilians." While there may be examples that imply there is still a possibility in FB-H to think balkanization in terms of coexistence, even if it is hanging on by only a thread, the divisions within the city are ever-present. On an economic level, to go by taxi from Eastern Sarajevo (governed by the Serbs) into Sarajevo (governed by Bosniaks and Croats), taxi drivers need to remove the taxi sign. It is against the law for a taxi registered in Eastern Sarajevo to pick up a passenger in Sarajevo that belongs to the Federation, and vice versa. Segregation is monetized.

The question that is pertinent for all unreconstructed buildings and cities is not whether they should remain in their ruinous state, but the role they may play in constructing the future post-socialist identity. In other words, how to deal with ruins and post-socialism via Balkanism in a way that does not reinforce the already violent association of the discourse. For Vukovar and Sarajevo, the identity is largely tied to memories of recent war, while for Mostar it is one of (falsely) constructed and spectacle-oriented coexistence and connection. Dubrovnik is straddling its identification with war and its mercantile history. Thinking trauma and reconstruction simultaneously needs to be counterposed against the thinking on how to apply balkanization in contemporary times without imposing borders and homogeneity. In other words, the question of balkanization is how to coexist with other beliefs and practices, and beyond the normative instatements of historical meta-narratives to do with land rights and ethnic homogeneity. One key agenda in any aftermath reconstruction is that the suffering and violence of the past are not to be brought into the present for purposes of fuelling animosity and eliminating the possibility to engage in more than one history.

Notes

1 Mient Jan Faber, "Cold Wars and Frozen Conflicts: The European Experience," in *Global Insecurity*, ed. Mary Kaldor (London: Continuum, 2000), 86.
2 Susan Woodward, *Balkan Tragedy: Chaos and Dissolution After the Cold War* (Washington, DC: Brookings Institution, 1995), 169.
3 Faber, "Cold Wars and Frozen Conflicts," 86.
4 Faber, "Cold Wars and Frozen Conflicts," 86.
5 Wilmer, *The Social Construction of Man, the State, and War*, 82.
6 Andrej Grubačić, *Don't Mourn, Balkanize!* (Oakland: PM Press, 2010).
7 V. P. (Chip) Gagnon Jr., "Yugoslavia in 1989 and After," *Nationalities Papers* 38:1 (2010): 29.
8 Zlatko Dizdarevic, *Sarajevo: A War Journal* (New York: Fromm International, 1997).
9 Branka Magaš, *The Destruction of Yugoslavia: Tracking the Break-Up 1980–92* (London: Verso, 1993), 179–218.

10 Leslie Benson, *Yugoslavia: A Concise History* (Basingstoke: Palgrave Macmillan, 2004), 105.
11 Vassilis Fouskas and Bülent Gökay, *The New American Imperialism: Bush's War on Terror and Blood for Oil* (Westport: Greenwood Political Publishing, 2005), 116.
12 Woodward, *Balkan Tragedy*, 198.
13 From the post–Cold War era, the first recorded use of the term urbicide occurred with the dissolution of Yugoslavia; the term urbicide became conflated with actual war, and an outcome of ethnonationalist violence, and thus an aspect of negating difference. In other words, the urbicide of cities in the disintegrating SFRY started to inflect the Cold War concept of urbicide as it became associated with war, the military, economics, mass media and law. The first inflected use of the term was by Mostar Architects Association in 1992. For more information about the Association's use of the term specifically, and more generally in terms of the dissolution of Yugoslavia and other parts of the world, see: Mostar Architects Association, "Mostar '92 – Urbicide," *Spazio e Societa`-Space and Society* 16:62 (1993); Bogdan Bogdanović, "The City and Death," in *Balkan Blues: Writing Out of Yugoslavia*, ed. Joanna Labon (Evanston: Northwestern University Press, 1995), 37–74; Ejub Štitkovac, "Croatia: The First War," in *Burn This House: The Making and Unmaking of Yugoslavia*, ed. Jasminka Udovicki and James Ridgeway (Durham: Duke University Press, 1997); Andrew Herscher, "Warchitectural Theory," *Journal of Architectural Education* 61:3 (2008); Robert Bevan, *The Destruction of Memory – Architecture at War* (London: Reaktion Books, 2007); James Raven, "Introduction: The Resonances of Loss," in *Lost Libraries: The Destruction of Great Book Collections Since Antiquity*, ed. James Raven (New York: Palgrave Macmillan, 2004); Andrew Herscher, "Urbicide, Urbanism and Urban destruction in Kosovo," *Theory & Event* 10:2 (2007); Martin Shaw, "New Wars of the City: Relationships of 'Urbicide' and 'Genocide,'" in *Cities, War and Terrorism: Towards an Urban Geopolitics*, ed. Stephen Graham (Oxford: Blackwell, 2004); Stephen Graham, "Lessons in Urbicide," *New Left Review* 19 (2003); Stephen Graham, "Constructing Urbicide by Bulldozer in the Occupied Territories," in *Cities, War and Terrorism: Towards an Urban Geopolitics*, ed. Stephen Graham (Oxford: Blackwell Publishing, 2004); Martin Coward, "Urbicide in Bosnia," in *Cities, War and Terrorism*, ed. Stephen Graham (Oxford: Blackwell, 2004), 154–172; Cynthia Simmons, "Urbicide and the Myth of Sarajevo," *Partisan Reviews* 68:4 (2001): 624–30; Martin Coward, *Urbicide: The Politics of Urban Destruction* (New York: Routledge, 2009); Martin Coward, "'Urbicide Reconsidered," *Theory & Event* 10:2 (2007); Martin Coward, "Against Anthropocentrism: The Destruction of the Built Environment as a Distinct Form of Political Violence," *Review of International Studies* 31:3 (July 2006): 419–437; Michael Q. Dudley, "Revisiting Cold War Ideology in the Secure City: Towards a Political Economy of Urbicide," *Theory & Event* 10:2 (2007); Sara Fregonese, "The Urbicide of Beirut? Geopolitics and the Built Environment in the Early Phases of the Lebanese Civil War (1975–76)," *Political Geography* 28:5 (2009): 309–318; Kanishka Goonewardena, "Marxism and Everyday Life: On Henri Lefebvre, Guy Debord and Some Others," in *Space, Difference, Everyday Life: Reading Henri Lefebvre*, ed. Kanishka Goonewardena et al. (London: Routledge, 2008), 117–133; Sara Hanafi, "Explaining Spacio-cide in the Palestinian Territory: Colonization, Separation, and State of Exception," *Current Sociology* 61:2 (2012): 190–205; Kenneth Hewitt, "Proving Grounds of Urbicide: Civil and Urban Perspectives on the Bombing of Capital Cities," *ACME: An International e-Journal for Critical Geographies* 8:2 (2009): 340–375; Adam Ramadan, "Destroying Nahr el-Bared: Sovereignty and Urbicide in the Space of Exception," *Political Geography* 28:3 (2009): 153–163; Michael Safier, "Confronting 'Urbicide': Crimes Against Humanity, Civility and Diversity and the Case for a Civic-Cosmopolitan Response to the Attack on New York," *City* 5 (2001): 416–429; David Campbell et al., "Urbicide: The Killing of Cities," *Theory & Event* 10:2 (2007); Bruce Stanley, "The

City-Logic of Resistance: Subverting Urbicide in the Middle East City," *Journal of Peacekeeping and Development* 27:3 (2013): 10–24; Iain Watson, "(Re) Constructing a World City: Urbicide in Global Korea," *Globalizations* 10:2 (2013): 309–325; Nurhan Abujidi and Han Verschure, "Military Occupation as Urbicide by 'Construction and Destruction': The Case of Nablus, Palestine," *The Arab World Geographer/Le Géographe du monde arabe* 9:2 (2006): 126–154; Nicholas Adams, "Architecture as the Target," *Journal of the Society of Architectural Historians* 52 (1993): 389–390.

14 Briefly, the first recorded use of the term urbicide during the Cold War era occurred in the 1960s. It first appeared in passing in the 1963 science fiction short story "Dead God's Homecoming." The second use was in 1966 to comment on the 1968–72 erection of the New York World Trade Centre (WTC). The third use was to describe the destruction of the Amoskeag mill complex in Manchester (New Hampshire) that began in 1968. The fourth use of the term was in 1987 to describe the 1950s–1970s "regeneration" of the South Bronx borough of New York. For more information on Cold War understanding of urbicide see: Michael Moorcock, "Dead God's Homecoming," in *Stormbringer* (London: Millenium, 1993); Wolf Von Eckardt, "New York's Trade Center: World's Tallest Fiasco," *Harper's Magazine* (1966); Leonard I. Ruchelman, *The World Trade Center – Politics and Policies of Skyscraper Development* (New York: Syracuse University Press, 1977); Ada Louise Huxtable, *Kicked a Building Lately?* (New York: Quadrangle/The New York Times Book Co., 1976); Ada Louise Huxtable, "Manchester, N.H.: Lessons in Urbicide," in *Goodbye History, Hello Hamburger: An Anthology of Architectural Delights and Disasters*, ed. Ada Louise Huxtable (Washington, DC: Preservation Press, 1986); Tamara K. Hareven, *Amoskeag: Life and Work in an American Factory-City* (Hanover, NH: University Press of New England, 1978); Marshall Berman, "Among the Ruins" *New Internationalist* 178, December 1987; Marshall Berman, "Falling Towers: City Life After Urbicide," in *Geography and Identity: Living and Exploring Geopolitics of Identity*, ed. Dennis Crow (Washington, DC: Maisonneuve Press, 1996).

15 Bogdanović, "The City and Death," 53.

16 Bogdanović, "The City and Death," 37–62. Martin Coward makes a similar point in his book *Urbicide: The Politics of Urban Destruction*. There, he identifies that "distinctive feature of urbanity to be "heterogeneity" ' (refer to page 49). Also, Coward writes that "[t]he logic of urbicide is one in which a number of events coalesce into a distinct pattern of destruction, the meaning of which is the destruction of heterogeneity in and through the destruction of buildings" (refer to page 53). However, unlike Bogdanović's differentiation of urban vs. non-urban conflict, for Coward "it is the built environment that is the target, not any specific form of urban life" (refer to page 51).

17 Xavier Bougarel, "Yugoslav Wars: The 'Revenge of the Countryside' Between Sociological Reality and Nationalist Myth," *East European Quarterly* 33:2 (June 1999): 157–175.

18 UNESCO, "Report on the Status of Cultural Heritage in the Old Town of Dubrovnik Following the Bombardments of October, November and December 1991," in *Report of the UNESCO Mission to Dubrovnik 27 November – 22 December 1991*, 22.

19 UNESCO, "Report on the Status of Cultural Heritage in the Old Town of Dubrovnik," 22.

20 UNESCO, "Report on the Status of Cultural Heritage in the Old Town of Dubrovnik," 27.

21 Stitkovac, "Croatia: The First War," 155.

22 Wilmer, *The Social Construction of Man, the State, and War*, 116.

23 The term 'constituent peoples' was used to describe the main nations in Yugoslavia: Croats and Slovenians (Roman Catholicism brought by Charles the Great – the Holy Roman Emperor, and the Austrian Habsburgs); the Muslims-nationality (Islam, brought by the Ottomans); Macedonians, Montenegrins and Serbs (Eastern Orthodox Christianity

brought by the Byzantine Empire). The constituent peoples could not be considered a minority and during Yugoslavia, could not declare autonomy in other republics, for example, Serbs in Croatia. The non-constituent peoples were national minorities who comprised over 26 ethnic groups. These included Slovaks, Hungarians, Romanians, Albanians, Italians, Romani and Czechs. The constituent and non-constituent peoples could also define themselves with the Yugoslav designation, representative of the whole country. Where non-constituent nations were a notable majority, such as Slovaks and Hungarians in the Autonomous Province of Vojvodina or Albanians in the Autonomous Province of Kosovo-Metohija (located in the Socialist Republic of Serbia), the minorities could – amongst other rights – speak in their own language. The education system also accommodated language differentiations, and street signs in noted areas were displayed in both languages; for example, Serbo-Croatian and Slovak in Vojvodina.

24 Kaplan, *Balkan Ghosts*; Misha Glenny, *The Fall of Yugoslavia* (London: Penguin, 1996).

25 Cited in David N. Gibbs, *First Do No Harm: Humanitarian Intervention and the Destruction of Yugoslavia* (Nashville: Vanderbilt University Press, 2009), 87; Glenny, *The Fall of Yugoslavia*, 12–14.

26 Gibbs, *First Do No Harm*, 87.

27 Wilmer, *The Social Construction of Man, the State, and War*, 136–137.

28 Bob Allen, "Why Kosovo? The Anatomy of a Needless War," *Canadian Centre for Policy Alternatives – BC Office* (July 1999): 11.

29 Woodward, *Balkan Tragedy*, 1; Gibbs, *First Do No Harm*, 163, 215.

30 Jon Calame and Esther Charlesworth, *Divided Cities* (Philadelphia, PA: University of Pennsylvania Press, 2009), 103–120; Bevan, *The Destruction of Memory – Architecture at War*, 9–10.

31 Coward, *Urbicide: The Politics of Urban Destruction*, 6.

32 Mostar Architects Association, "Mostar '92 – Urbicide," *Spazio e Societa`-Space and Society* 16:62 (1993).

33 Mostar Architects Association, "Mostar '92 – Urbicide," 9.

34 Mostar Architects Association, "Mostar '92 – Urbicide," 19.

35 Herscher, "Warchitectural Theory," 37.

36 Herscher, "Warchitectural Theory," 37. Association of Architects DAS – SABIH (Asocijacija arhitekta DAS – SABIH), *Warchitecture – Urbicide Sarajevo/Sarajevo, Une Ville Blessée*, Bordeaux, 1994.

37 Cited in András J. Riedlmayer, "From the Ashes: The Past and Future of Bosnia's Cultural Heritage," in *Islam and Bosnia: Conflict Resolution and Foreign Policy in Multi-Ethnic States*, ed. Maya Shatzmiller (Montreal: McGill-Queens University Press, 2002), 105.

38 Cited in Edgardo Civallero, "When Memory Turns into Ashes . . . Memoricide During the XX Century," *Information for Social Change* Summer (2007): 2.

39 Coward, *Urbicide: The Politics of Urban Destruction*, 7–8.

40 Ministry of Defence of the Republic of Croatia, "Minister Krstičević Meets with the U.S. Secretary of Defense – Press Release," 13 July 2017, < www.morh.hr/en/news/press-releases/15027-minister-krsti%C4%8Devi%C4%87-meets-with-the-u-s-secretary-of-defense.html>

41 Faber, "Cold Wars and Frozen Conflicts," 92.

42 Paul Latawski and Martin A. Smith, *The Kosovo Crisis and the Evolution of Post-Cold War European Security* (Manchester: Manchester University Press, 2003), 82.

43 Bogdanović, "The City and Death," 53.

44 Bogdanović, "The City and Death," 54.

45 Bogdanović, "The City and Death," 47–48.

46 Bogdanović, "The City and Death," 47–48.

47 Despite Bogdanović dying in 2010 – 15 years after the full dissolution of Yugoslavia, and 11 years after NATO's 1999 targeting – his initial writings on urbicide did extend past his thoughts laid out in the "The City and Death," where he largely dealt with the destruction.

48 The history of Vukovar also includes the immediacy of the Vucedol archeological site, which points to this area being the early settlements of Troy (3,000–2,500 BC) as well as showcasing the patterns of living of the first Indo-Europeans who passed through and lived in these areas. See: Britt Baillie, "Capturing Facades in 'Conflict-Time': Structural Violence and the (Re)construction Vukovar's Churches," *Space and Polity* 17:3 (2013): 301.

49 Benson, *Yugoslavia: A Concise History*, 35.

50 See: Catherine Baker, "Music as a Weapon of Ethnopolitical Violence and Conflict: Processes of Ethnic Separation During and After the Break-Up of Yugoslavia," *Patterns of Prejudice* 47:4–5 (2013): 409–429; Wolfgang Höpken, "Between Civic Identity and Nationalism: History Textbooks in East-Central and Southeastern Europe," in *Democratic Transition in Croatia: Value Transformation, Education and Media*, ed. Sabrina P. Ramet and Davorka Matić (Austin: Texas A&M University Press, 2007), 163–192; Kruno Kardov, "Remember Vukovar: Memory, Sense of Place and the National Tradition," in J. Russell and R. Cohn, *Battle of Vukovar* (Edinburgh: Bookvika, 2012).

51 Britt Baillie, "Memorialising the 'Martyred City': Negotiating Vukovar's Wartime Past," in *Locating Urban Conflicts*, ed. Wendy Pullen and Britt Baillie (London: Palgrave Macmillan, 2013), 115–131.

52 Council of Europe Committee of Ministers, "Recommendation Rec (2001) 15 of the Committee of Ministers to member states on history teaching in twenty-first-century Europe" (Adopted by the Committee of Ministers on 31 October 2001).

53 Antonio Aguilera, "Ethics and Violence," in *Globalization, Political Violence and Translation*, ed. Esperanza Bielsa and Christopher Hughes (Basingstoke: Palgrave Macmillan, 2009), 79.

54 Britt Baillie, "Vukovar's Divided Memory: The Reification of Ethnicity Through Memorialisation," in *Divided Cities/Contested States Working Paper Series/Working Paper 25 – Conflict in Cities and the Contested State 2007–2012*, 10.

55 Refer to: Victor Alneng, "The Modern Does Not Cater for Natives: Travel Ethnography and the Conventions of Form," *Tourist Studies* 2:2 (2002): 119–142; Victor Alneng, " 'What the Fuck Is a Vietnam?': Touristic Phantasms and the Popcolonization of (the) Vietnam (War)," *Critique of Anthropology* 22:4 (2002): 461–489; Maureen Moynagh, *Political Tourism and Its Texts* (Toronto: Toronto University Press, 2008); Michael Pretes, "Tourism and Nationalism," *Annals of Tourism Research* 30:1 (2003): 124–142.

56 Patrick Naef, "Tourism and the 'Martyred City': Memorializing War in the Former Yugoslavia," *Journal of Tourism and Cultural Change* 14:3 (2016): 222–239.

57 It needs to be mentioned that the unveiling of the monument in Zagreb was met with protects by a number of organizations, some of which include the Youth Initiative for Human Rights and the Anti-Fascist League of the Republic of Croatia.

58 Vladimir Dedijer, *The Yugoslav Auschwitz and the Vatican: The Croatian Massacre of the Serbs During World War II* (London: Prometheus Books, 1992).

59 Rory Yeomans, "How to Combat Croatian Revisionists' Culture of Lies," *Balkan-Insight*, 6 June 2018, <www.balkaninsight.com/en/article/how-to-combat-croatian-revisionists-culture-of-lies-06-05-2018>

60 Yeomans, "How to Combat Croatian Revisionists' Culture of Lies." Also see Rory Yeomans, *Visions of Annihilation* (University of Pittsburgh Press, 2013); and Rory Yeomans, *The Utopia of Terror* (Rochester: University of Rochester Press, 2015).

61 Kolstø, " 'Western Balkans' as the New Balkans," 1255.

62 Susan Forde, "The Bridge on the Neretva: Stari Most as a Stage of Memory in Post-Conflict Mostar, Bosnia-Herzegovina," *Cooperation and Conflict* 51:4 (2016): 476.
63 Forde, "The Bridge on the Neretva," 477.
64 Georg Simmel, "Part I Modernism – George Simmel," in *Rethinking Architecture*, ed. Neil Leach (London: Routledge, 1997), 64.
65 Kristen M. Hartmann, "Fragmentation and Forgetting: Sarajevo's Vijećnica," *International Journal of Heritage Studies* 22:4 (2016): 320.
66 Naef, "Tourism and the 'Martyred City,'" 2016.
67 This quote appears on one of the photos in the corridor of the Museum of Alija Izetbegovic.
68 This quote appears on one of the photos in the corridor of the Museum of Alija Izetbegovic.
69 Guy Robinson and Alma Pobrić, "Nationalism and Identity in Post-Dayton Accords: Bosnia-Herzegovina," *Journal of Economic and Social Geography* 97:3 (2006): 239.

4 Allied operations

Present-future partnerships of humanitarianism, peace and victory

Rhetoric, destruction and Operation Allied Force

Humanitarianism and its paradoxes

To think of Balkanism as the peripheral and liminal Other has an opportunist element: to use Balkanism as a means to deal with questions to do with violence constructed, distanced and airbrushed nationally, regionally and by Anglo-American and other Western countries. It is, in other words, a process by which certain acts of violence are normalized and/or ennobled as humanitarian, while others are constructed as acts of deep-seated barbarism. The premise of using Balkanism will be to consider violence in its forms and limitations, particularly when the identification of Balkanism is made to rely on parameters of perennial war and transient peace. Thus, the liminal space of Balkanism and its overlapping with a fragmentary dynamic of balkanization will be deployed to start deconstructing this binary setup.

The concept of violence has varying interpretations, depending upon the audience and the political agenda. One that stands out in its peculiarity is when it is exercised for legitimate and humanitarian reasons, deployed for purposes of eliminating violence, but through the barrels of guns. The argument will be that taking aim with a humanitarian agenda not only purifies violence, but also exonerates the infliction of violence when serving justice is the cause. In fact, it perpetrates a play on words, considering that humanitarianism is not readily associated with acts of war and violence; indeed, "waging war to prevent war appears to be as farcical as fucking for virginity."[1]

The argument in international relations may be that in cases of severe breaches of human rights, the inviolability of sovereign rights is suspended and concerned states have a moral responsibility to intervene.[2] The track record of applying this logic seems to have been selective, as seen in violence supported by the US in Congo in late 1990s; clearly, these zones were not seen as deserving of humanitarian 'justice.' The question similarly pertains as to why the US and/or NATO and/or the UN did not intervene in Rwanda in 1994 to prevent some 800,000 people from being slaughtered. Selective evaluation of violence anticipates selective reasoning to distinguish the lives that need to be preserved from those that may be dispensed with. Humanitarian and moral zeal is seemingly mobilized relative only to specific ethnicities and areas of the world. NATO's intervention in Serbia to protect ethnic Albanians living in Kosovo calls into question its failure to intervene in Turkey (a NATO member) where ethnic Kurds have been massacred for years. There have been no attempts by NATO to restrain Turkey's armed forces. The massacres in Congo and Sierra Leone have also not been seen as examples of humanitarian crises. While pairs of historical examples are never fully parallel, the lack of consistency and proportionality in exercising humanitarianism indicates an absence of impartiality when committing to such operations.

NATO's foremost reason for launching Operation Allied Force was to end the violence and ethnic cleansing by the Milosevic regime in Kosovo.[3] On 31 March 1999, US President Bill Clinton stated that the objective of the Operation was to "raise the price of aggression to an unacceptably high level so that we can get back to talking peace and security, to substantially undermine the capacity of the Serbian government to wage war."[4] For the Secretary General of NATO, Javier Solana, the military Operation Allied Force was a necessary "humanitarian intervention"

> intended to support the political aims of the international community. It will be directed towards disrupting the violent attacks being committed by the Serb Army and Special Police Forces and weakening their ability to cause further humanitarian catastrophe [. . .] Let me be clear: NATO is not waging war against Yugoslavia. We have no quarrel with the people of Yugoslavia who for too long have been isolated in Europe because of the policies of their government. Our objective is to prevent more human suffering and more repression and violence against the civilian population of Kosovo. [. . .] We must stop an authoritarian regime from repressing its people in Europe at the end of the 20th century. We have a moral duty to do so. The responsibility is on our shoulders and we will fulfill it.[5]

One day after the targeting began, NATO's Defence Secretary George Robertson reaffirmed Solana's position by stating that the intervention was impelled by a need "to avert an impending humanitarian catastrophe by disrupting the violent attacks currently being carried out by the Yugoslav security forces against the Kosovar Albanians, and to limit their ability to conduct such repression in future."[6] The enactment of violence was possible, as it was described as a proportional and moral objective, reinforced in statements by Solana and Robertson that NATO would target significant Serbian military areas only. The intervention was required, as Serbia, seen as a barbaric and Balkanist regime in relation to both the treatment of Albanians in Kosovo and the 1990s acts of atrocity in Bosnia, was breaching the moral democratic values found within Europe and the Anglo-American parts of the world. These values could be defended only by military intervention.

The claim that the intervention was in the name of humanitarianism invoked the traditional associations of this notion: immediacy and emergency, mobilization to save lives and higher morals and values of humankind,[7] all of which legitimize the motivation and action behind humanitarianism. However, allowing a military organization such as NATO a free hand in defining and deciding when a violent situation warrants a humanitarian response, including articulating moral values, is of concern. Trust has been undermined in the past, such as when the rhetoric of humanitarian intervention was used during Mussolini's invasion of Ethiopia in the name of freeing slaves and introducing Western civilization, or when Hitler ordered the occupation of parts of former Czechoslovakia ostensibly to quell ethnic strife between Germans and Czechs.[8] Or, indeed, when the US army expelled Native Americans from their homes in the name of law and humanity.[9]

For the NATO operation and its mandate to retain ongoing support, both tight control of information through Orwellian doublespeak and a careful engineering of dehumanization were necessary; the latter made possible by observing the violence in Kosovo through the lens of the war in Bosnia and Herzegovina. This elastic rhetoric drew much of its credibility from the Balkans' liminal identification, which for Misha Glenny is evident from the British press' framing of the Balkans first as a region of primordial hatreds, and the very next moment as a zone of multicultural values and the very pulse of Europe, as epitomized by Sarajevo.[10] Needless to say, the elastic reasoning for the intervention, including associations with Sarajevo, starts to place a clear measure and ordering on the violence occurring in Kosovo.

With Kosovo viewed through Bosnia's lens, there was an undercutting of partiality and objectivity. By largely pointing a finger at Milosevic as the reason for the intervention, and (in)directly for the outcome of the intervention in terms of overall damage, the complexity of the violence was also reduced. While Milosevic's regime was certainly a cause for concern given its nationalist and violent tendencies, the visibility of violence enacted by NATO was significantly obscured as its enactment placed neither personnel nor equipment on the ground. Instead, it was fought remotely from the air and through the infrastructure of satellites and physical wires. The distancing was paralleled by a slippery narration and perception of Milosevic. While the Western mass media largely condemned his positions, and despite the West supporting the dissolution of Yugoslavia, Milosevic for a large part of the late 1980s and 1990s was seen as "a man we can do business with," particularly when it came to the Dayton agreement.[11]

Spectacle, distance and perception management

Managing the audience's perceptions about the people in Serbia and Kosovo and the events taking place there was possible because the Kosovo conflict was the first war fought from the air alone. Even when civilian deaths did occur, NATO simply acknowledged that infrastructure was hit.[12] The focused framing of violence is demonstrated in NATO's 'accidental' targeting by two projectiles at 11:40 of a civilian-occupied Express Belgrade-Nis-Skopje train 393 crossing the Grdelica railway single track bridge. The bridge is located 250 kilometres southeast of Belgrade, on a transport route that, according to NATO, was an important line of supply for the Serbian paramilitary forces in Kosovo.[13] This targeting occurred at 11:40 on 12 April 1999, Orthodox Easter Monday.

According to NATO spokesman Jamie Shea, "a very extensive analysis [. . . was done] which shows the pilot was totally unable to realize, to know before releasing his weapon that a train would appear on the bridge."[14] The narrative of the supreme commander of NATO General Wesley Clark is even more detailed in maintaining that the pilot was focused on the bridge,

> when all of a sudden, at the very last instant, with less than a second to go, he caught a flash of movement that came into a screen and it was the train coming in. Unfortunately, he couldn't dump the bomb at that point. It was locked, it was going into the target and it was an unfortunate incident which he and the crew and all of us very much regret.[15]

The narrative was that the train was travelling too fast for the trajectory of the missiles to have been changed in time to avoid the killing of 14 civilian passengers and wounding 60 more (some of whom were children).[16] The video was replayed on Western television screens incessantly in order to demonstrate that the speed of missiles means that 'accidents' such as these are often unavoidable.[17] The targeting was seen as violent only when a train emerged and became visible on the bridge. If this information had remained undisclosed and unseen, it would have evaded the classification of being violent. This was a case not only of audience emotions being managed through control of media representation, but also fear of those watching the spectacle from a distance that if support is not generated for the elimination of the Other, what is happening in Other zones may very well come knocking on Western doors.

The disciplining is not overt. It is a result of the instilled fear that, if war is not waged, security in the 'civilized' Western world is likely to be breached. The paradox of upholding civilized and humane values is that they are implemented by creating borders to protect from the Other whilst simultaneously targeting the Other. To think in a civilized way would mean the elimination of obstacles to think and act differently from the norm. Yet the question is whether this type of civilization is feared, since exercising it in such a way would mean that one is no longer driven to inscribe power and control on the Other.

In January 2000, the unavoidability of that incident was called into question. The footage shown at the initial NATO press conference was revealed to have been altered by showing it at three times the actual speed recorded by a camera installed in the warhead of one of the missiles targeting the bridge.[18] The discrepancy in speed was attributed to technology; Shea was noted as saying that the speeded-up video was caused by a "technical phenomenon" rather than human manipulation.[19] Using technology to deny responsibility for atrocities was also seen after WWII. According to Virilio, Albert Speer used the efficiency of technology in his defence during the Nuremburg Trial in order "to prove that he was only an instrument, certainly guilty, but that technological advances, in particular in the field of communications, had issued in the catastrophe."[20] The argument here is not that NATO's 'accidental' killing of 14 civilians is comparable to the large-scale genocide of Jews during WWII. Instead, the comparison suggests that technology is now an instrument that can erase evidence of (possibly deliberate) crime through a process of speed-in-variation, and it reminds us that instrumentation can be framed as a perpetrator of crime. Technowar is not replacing human involvement and impact. It is simply that the speed and multipurpose use of missiles can be deployed for erasing evidence of (a possible) crime. The development of technology, changes in military strategies, and public culture have all had an impact on media and war reporting.

NATO's 1999 Operation reveals that claiming to act on behalf of humanitarianism is nothing more than a disguise for a new wave of colonialism. The conditions under which supposedly humanitarian motives were implemented were those of the discourse of Balkanism. The rogue Balkanist agenda was persistently used during the Operation, as evident in the speech by the British Prime Minister Tony Blair on 17 May of that year when he stated that the world could not "stand by when a rogue state brutally abuses the basic rights of those it governs."[21] For the Balkanist rhetoric to stick, violence needed to be monitored and framed via an infrastructure of circuitry and satellites, allowing NATO missiles equipped with a camera to at once destroy, survey and record the targeted infrastructure on the ground. Technology was not only deployed to represent the absence and presence of violence through speed-in-variation. It was also an attempt to first miniaturize the weapon and then defamiliarize the military frame through a camera lens that could eliminate the human (subjective) gaze and replace it with a technological (objective) gaze. Thus, Cruise Missiles became humanitarian machines and weapons of democracy, frames of spectacle and tools of legal evidence, where any 'gaps' in erasing and constructing a narrative could be seamlessly sutured. Here, the perception of a particular zone is through a limited window and vision of a media spectacle.

The Operation exceeded the "humanitarian" dictum, since the disproportional aspect of the Operation – NATO air power versus unarmed civilians – was, for Andrew Herscher, "destruction without humanity."[22] Herscher's "destruction without humanity" is evident in the transition of NATO's "humanitarian intervention" from a pretext of saving Albanian lives to an outcome of adding to the overall Albanian death tally. The 14 April bombing of a convoy of Albanian refugees on a Prizren-Djakovica road killed 75 civilians and wounded over 60; a month later, another 50 Albanians were killed in Korisa. If proportional strikes are synonymous with the weighing of the military target in terms of the harm the strike may cause, or the prevention and minimization of civilian casualties, the point at which the permissible limit of violence is exceeded becomes open to interpretation.

The disproportionality of Operation Allied Force is made obvious in its claim of "no casualties on our [NATO's] side" (the first war able to boast this) even as it killed 500 civilians on the ground.[23] Only numbers of civilian deaths are highlighted because, according to Eyal Weizman, the killing of civilians has now turned into a "risk transfer war" where what is considered a proportional killing of civilians in relation to combat fatalities is dependent on what territory one is speaking about and who is doing the calculations.[24] The paradox of proportional violence is thus that it has been presented on one hand as "a mathematical-minimum problem," while on the other the formula one relies on is always subsumed in "the economy of variations."[25] The mathematical calculation not only optimizes violence but also sets up the possibility for operative and efficient violence, all of which are permissible in control societies.[26]

Considering that the Balkanist corpse seemingly has less value than that of an Anglo-American or other Western, the ability to discern the measure and distribution of excessive violence, including the grievability of certain populations, is eliminated from the start. If lives are ungrievable and zones are marked as barbaric, as Judith Butler tells us, there is nothing to mourn; alternatively, any destruction incurred is insignificant since both the zones and their people never existed as they did not fit within the Western and Anglo-American value systems, particularly when an intervention is undertaken for purposes of defending security and peace.[27] With the bombing of the Federal Republic of Yugoslavia (FRY),[28] US President Bill Clinton's rhetoric was that "we are upholding our values, protecting our interests, and advancing the cause of peace,"[29] which might well prompt speculation that the lack of democratic practices in the US is proportional to its drive to delivery democracy to the rest of the world.[30] The way in which the audience is recruited for such spectacles is through images and rhetoric generated from a distance, justified on moral humanitarian grounds built on the logic of violence inflicted on the Other as a way to regain and fulfil a sense of self.[31] Or, as Elaine Scarry would put it, "[t]he human capacity to injure other people has always been greater than its ability to imagine other people."[32]

Digital technology permits war to be waged over greater distances as well as shrinks our ability to imagine what the Other may be experiencing, particularly when the objective or person is rendered a legitimate target under the proportionality dictum. However, virtual wars accommodated by digital technology and satellites are virtual only on one side; those experiencing them on the ground certainly feel the effects of violence. Yet we live in a time when humanity is framed through the lens of progress and in terms of instatement of security on grounds of fear of the Other. The belief in such rhetoric is not just a matter of propaganda or censorship; it reflects the perversity of the age we live in when the Other is framed through the lens of a second-rate citizen, certainly possible when the derogatory Balkanist rhetoric is used. Likewise, when the life or death of the Other has no value, the value and potential growth of the self is also limited. The virtual wars rob us of emotion and empathy largely because distance facilitates a limited framing of the events taking place on the ground.

Reporting from a distance, for Jean Baudrillard, turns its audience into hostages of information.[33] It is not that he is likening the audience to actual hostages, but that he is putting focus on the information media and the way in which they serve to rob the audience of the diversity of information narratives taking place around the world. Through carefully constructed images, the viewers of news stories are habituated to perceive the violence taking place in the world through the gaze of a military-industrial-media-entertainment complex (MIME).[34] In contrast to the wars of the ancients, which were sporadic and where possession of forts and citadels was contested in the fairly transparent language of war-fighting techniques, or with modern, more systematic and centralized wars, current methods of warfare are less overt, and possibly more dangerous, as their visibility is becoming more opaque; they are now fought from labs and media hubs.[35] One can represent it as a highly computerized and digitized war assault, the last intervention of the pre-drone era.

Technowar and erasure of violence

Technowar is an opportunity to not only exploit the capacities of smart machines but also to eliminate acts of violence. While a measure of technological and media exploitation did occur in the military campaign of the 1991 Gulf intervention that made it hard to tell what did and did not happen as the narrative and the understanding about the war took place more on television,[36] Kosovo was the first internet war instrumentalized from air power alone. It was a war of the airwaves, where not only was information gathered but the military operation itself was executed from a distance, distance facilitating control of the act. Here, there is a management of how the targeting is framed and how history is constructed and managed. MIME is instrumental in this administration, which is cast in the present-continuous in that there is a disengagement from the issue of humanitarian motives and their credibility. After all, humanitarianism is motivated by bringing peace and democracy rather than inflicting violence and war.

These days, the battle has no start or end in that it infiltrates into every aspect of the ether. It is an all-encompassing battle made possible with the simulation of media and the military in order to discipline. It exists in the militarization of information presented on media and television screens. This is an age of information and data, where the focus is not necessarily on the control of physical power but on the control of information.[37] The data generated by technological devices such as predators, which are equipped with tiny cameras, and were originally designed for purposes of surveillance; now they are also used as weapons in that they screen the context. These devices were used for the first time in the 1990s wars in the Balkans.[38] Balkanization in this instance is deployed for purposes of grouping data, securitizing it, as well as managing which data has to be processed and transmitted to the public to perpetuate the uniform belief in an intervention perpetrated on behalf of humanitarian and democratic values.

That information control was an important NATO objective was seen in the targeting of the government-owned and pro-Milosevic Radio Television of Serbia (RTS) building on the night of 22–23 April 1999. The importance of establishing a uniform rhetoric was also present during Milosevic's rule. Up until 1998, privately owned media such as radio B92 and newspapers such as *Dnevni telegraf* (*Daily Telegraph*) operated alongside the pro-Milosevic RTS, meaning that pro-Western views – including an open critique of the situation in Kosovo – were permitted and consumed.[39] However, in October 1998 – five months prior to NATO's intervention – media legislation was passed at the federal level imposing heavy fines on any media outlet that was openly against Milosevic.[40] The law was used to crack down several times on *Dnevni telegraf*, which resulted in other newspapers toning down their anti-Milosevic stance or suspending their publication altogether.[41]

The stance of privately owned media was in complete contrast to that of the RTS, which neglected to mention that Serbian military operations did involve attacks on Albanian civilians, sometimes targeting whole villages irrespective of whether civilian residents were real or imagined supporters of the Kosovo Liberation Army (KLA).[42] In that context, the RTS building was seen as a legitimate NATO target. This is despite General Wesley Clark stating that NATO

> knew when we struck that there would be alternate means of getting the Serb Television. There's no single switch to turn off everything but we thought it was a good move to strike it and the political leadership agreed with us.[43]

The attack on a building, while 120 technical and production staff were working there, disrupted Serbian TV broadcasts for three hours at the cost of 16 lives and 18 wounded. NATO's Operation seems also to have been an attempt to establish the hegemony of a single market of information, which paradoxically is an objective that NATO saw as undemocratic when legislated by Milosevic and pursued through his RTS media. The liability for lives lost during the targeting of RTS was disputed; the only compensation offered for such violations perpetrated by NATO was to the Chinese government for the 'accidental' destruction of its embassy building in Belgrade some two weeks later. This is despite schools, hospitals, museums and even the territory of Bulgaria (inadvertently) being bombed. Moreover, the payment entailed no acknowledgement of legal liability.[44]

In terms of NATO's targeting and framing of events, the question to do with proportionality and territoriality is no longer only reminiscent of Foucault's thinking about power, in terms of analysing and breaking down spaces, places, individuals and operations into components ready to be modified.[45] In terms of NATO's 1999 Operation, both Western and Serbian media made a clear identification of who were the victims and who were the perpetrators during the violence in Kosovo. Questions to do with power are also no longer just about the alteration of NATO's list of proportional military targets, which within a very short period following the start of its targeting campaign included cultural and social centres – the whole city was turned into a target.

Today, years after the 1999 Operation, proportionality is seen in entrenched progressive training and perpetuated control via Serbia's cooperation with NATO. Serbia has agreed to NATO's Individual Partnership Action Plan (IPAP), which ensures "that Serbian military personnel are able to work effectively and safely within the UN and EU missions in which they serve" and "[that Serbia] develop the capacity of its forces to participate in UN-mandated multinational operations and EU crisis management operations. These are areas in which NATO and individual Allies have much expertise to offer."[46] One of these 'expert' operations was undoubtedly NATO's Operation Allied Force in 1999. Despite these developments, the Serbian President (since May 2017) Aleksandar Vucic has repeatedly been reported in Serbian media as insisting that Serbia will remain militarily neutral and will not join NATO. The business end of political deceit seems to be paramount in this media construct.

In terms of NATO's broad-scale use of military forces and territorial management, it is no longer only a reflection of Foucault's thinking on governability filtered through various institutions, procedures and tactics, as seen in Serbia's adoption of NATO's military training techniques. Instead, this is a question of, firstly, eliminating all spaces that operate with an alternative structure such as Titoist or Balkanist and, secondly, making provisions for this city and region to become absorbed by the 'security' provisions and in the name of 'protection' by extensively becoming subsumed into the NATO protectorate. Violence is a multi-faceted process whereby the territory of a country is first fragmented, blurred and paralyzed, then folded in space and time to affect every facet of the society.

Managing the system of beliefs and values by conjuring the arts and culture was seen in US President Bill Clinton's speech. He used the speech, whose intent was to gain support for the 1999 intervention, to invoke Bob Dylan's 1960s "Blowin' in the Wind" lyrics by asking, "[h]ow many roads must a man walk down before you call him a man?" to imply that the air strikes were simply another attempt by America to intervene on behalf of freedom and human rights – albeit selectively by setting up the Kosovo Albanians as the victims and the Serbs as perpetrators. This is not just a question of performative rhetoric used to steer people's moral compasses in the direction deemed correct, but also a sign of a culture where it has become perfectly acceptable to misuse the lyrics of Bob Dylan, who is known for his critique of dominant power structures and the need to think counter practices. Such a misappropriation suggests that today we live not only in a culture of simulation but one of glazed stimulation. Baudrillard's thinking could not ring more true here: the real has been effaced by the signs of its existence, there is only the copy.[47] Nevertheless, the very premise of being able to critique suggests that the end-point of Baudrillard's theory has not yet been reached – which is not to say that it will not be in the near future.

That the media was a significant aspect of the intervention is also evident in the incorporation into the campaign of close monitoring of the media 24–7, with the dialogue between military, media and broader political aims geared towards achieving a singular narrative of Milosevic perceived as another Hitler, even if that narrative contradicted the rhetoric of the recent past in which Milosevic was a political figure with whom political agreement was possible. The start of NATO's targeting coincided with the mass flight of Albanians from Kosovo. According to the UNHCR, 4,000 Albanian refugees fled from Kosovo on 27 March, three days after the operation began.[48] By early June, the numbers surpassed the 850,000 mark.[49] In terms of Serb casualties, there were more in the first three days of the operation than in the three months prior to the 'humanitarian' intervention.

Seemingly, far from halting expulsion and violence experienced by Kosovo Albanians, a key driver for NATO's intervention, the 1999 targeting increased it.

History as referential violence

The media and military rhetoric during the intervention thus had to change from an objective of preventing the expulsion of Albanians by the Serbian paramilitary forces to one of acting to bring the Kosovo Albanian refugees home. In order to facilitate the shaping of a uniform – albeit elastic – rhetoric, the military and media dialogue extended to constant on-air briefings, interviews, press conferences, research, analysis and global media monitoring as well as to providing a platform for rebuttals.[50] Within this structure, language was used in media to occult NATO's role as the targeting agent. The language of reporting was one of passive voice, thus shifting the focus to events in Belgrade and to Serb violence, rather than on NATO inflicting damage.[51] When the agent of such damage was mentioned, it was framed with reference to inanimate objects such as bombs and aircraft, as opposed to animate agents such as NATO and pilots, thus further distancing any association between animate agents and crimes committed. The storyline continued the Hollywood blockbuster triptych scenario: clear perpetrator (Milosevic/Serbs), victim (Kosovo Albanians) and hero (NATO). This is a somewhat adjusted version of the triadic structure of 'ruler/victim/audience' seen in Foucault's writing on public torture, where "[a] successful public execution justified justice, in that it published the truth of the crime in the very body of the man to be executed."[52] To substantiate the justice in casting the Serbs and Milosevic on the barbarian side, reference was made to past crimes committed in Bosnia; it was a war of referential violence, via Srebrenica and Sarajevo.

The Serbian media did not play an innocent game either, though in its case, the ordering was different. The Serbs were collectively presented as heroes and innocent victims of NATO's targeting, while the role of the villain was attributed to NATO; NATO as a military organization rather than NATO as America or Great Britain as a nation, which is the key difference in the hero-victim-villain triptych.[53] In this instance, NATO was (in)directly associated with the Nazis since news of various targets hit by NATO would often be followed by reminders that Belgrade was also bombed by the Nazi Germans and by the Anglo-Americans during WWII. The narratives that populated the Serb media heavily relied on instances when NATO hit non-military targets; from schools and factories to hospitals and bridges. The focus was also on occasional demonstrations around the world protesting against NATO's targeting and in solidarity with the Serbian people,[54] coupled with elimination of any footage showing Serb violence directed towards the Albanians; the only voice given to Albanians was that of suffering as a result of NATO's targeting.[55]

Considering that the rhetoric of the operation was built upon shifting alibis, the images of destruction in Kosovo often required interpretation. During the 14 May 1999 briefing, NATO's spokesman Jamie Shea had to interpret evidence that buildings in Kosovo and Metohija had been destroyed by Serb forces.[56] According to Herscher,

> [t]he subjects of these images, pictured in a bitmapped haze of pixels, had to be enmeshed in a complex apparatus of written, graphic and verbal signs in order to become legible. The image of a building in flames, or damaged or even destroyed, was incomprehensible without an identifying text signifying flames, damage, or destruction. The rhetorical distance of irony in Shea's presentation thus mapped onto the spectatorial distance of his satellite imagery. Was it possible to get up close and personal to a satellite image in 1999? Was it possible to lead spectators from a network of pixels to a vale of tears?[57]

Shea's commentary may well be an example of what Weizman terms as only a criminal being able to interpret and solve a crime in that images are endowed with meanings and made to speak a rhetoric to suit the desired outcome.[58] The nature of these illegible images is peculiar considering that throughout the Operation, NATO used cameras mounted onto missiles to film the destruction of a particular target from the moment the weapon was discharged to the moment of the strike.[59] This is a different form of Weizman's sanitary rhetoric, where media in the Israeli-Palestinian conflict is deployed to set up a clear representation of a perpetrator and a victim. It is possibly more coercive still, as the law, military and media (seen in Shea's interpretation of pixels as evidence of destruction by Serb forces) form a trinity where the value of language and images is to *show less* through defamiliarization of a military frame.

The Western tradition derives knowledge from the faculty of sight. Restricting what is seen to a distanced view not only limits clarity in deciphering what is actually taking place on the ground, but also eliminates all other ways of looking at the event. The interpretation of objects of destruction and pixels of images no longer need be confined to legal proceedings but can be undertaken during press conferences and in the media generally. The politics of globalization is that history is inscribed in the 'live,' with media-military partnership playing a pivotal role in this inscription. That global news is still largely dominated by Anglo-American media, the output of which has been controlled by a small number of deregulated and privatized organizations, means that, despite there being more news networks, the raw material of reportage tends to come from a single source.[60] Similarly, the fact that media is driven by a commercial and profit-making motive means that it is not economically sustainable to source news independently from distant parts.[61]

The frame of image and story presented on the news constrains rather than exhibits reality, in particular that which needs to be seen as being real.[62] Containment well serves the desire to eliminate any alternative narrative. However, for the audience to participate in 'live' history, the theatre of spectacle needs to be staged in a way that makes images seem comprehensive, objective and intelligible yet also constructed to eliminate any questioning of the prevailing rhetoric. Considering that not only is 'live' history generally driven by the desire to organize the perception of the war,[63] but also that Anglo-American media is built upon the homogenization of rhetoric,[64] the ability to recall immediate and distant news is all the less possible since the objective is geared towards eliminating any questioning. The persistent underlying motive is to deliver the imagery of Other as the perpetrator. Despite the immediacy of images, the means and the skills to see have been largely lost.[65] The violence is present in that which has been removed and that which cannot longer be seen.

What history *in live* does, using Der Derian's thinking, is that it maps all that no longer exists.[66] After all, the Balkanist imaginary of Yugoslavia associates this border region with monstrosity, and as such, it is a place that cannot be comprehended by the civilized part of the world.[67] This is a history with no distance or access to percepts beyond what is shown on the screen; after all, the screen allows past, present and future to come together and collide in a dazzling array of action, color and narrative. That information becomes nothing short of a Hollywood movie in the making; the assemblage seen in the way in which the US managed to reidentify the KLA as well as Milosevic in the space of not much more than a year. Only a year prior to NATO's targeting, the US announced that it considered the KLA a terrorist group, "condemn[ing] very strongly terrorist activities in Kosovo."[68] In 1996–97, at the height of student protests against Milosevic's rigging of votes, the US described Milosevic as a democratic president. This is not to deny the atrocities by Serb forces in Kosovo, but it does bear out the broader certainty that villains and victims are created to suit the politics of the time. While the violence in Kosovo committed by the Serb paramilitaries, as well as by the KLA, was significant, the NATO 'humanitarian' intervention did not extinguish it, but rather increased it. Violence was not a side effect of the 'humanitarian' intervention directed towards the unruly Balkanist country, but its very method and policy.

Peace, prisons and justice: history in vacuum

Territory, peacekeeping and the Tower of Babel

With the Kosovo intervention made possible via virtual networks (from television and the internet to smart bombs and precision targeting), the finishing touches were implemented on the ground – once again with a visibility to be showcased on viewers' screen. Peace was brought to Kosovo on 3 June, after a 10-week targeting campaign. It took another six days for the strikes to terminate. The victory was marked by bringing in 55,000 NATO troops to Kosovo; the territory was balkanized under a number of measures claimed necessary in order to 'manage' peace. The first measure was driven by returns and expulsion of civilians. Within a month, according to the UNHCR, over 730,000 Kosovo Albanian refugees returned, whilst more than 170,000 people – predominantly Serbs and Romani – were expelled from Kosovo.[69] The second mode of balkanization was made possible by creating enclaves; those Serbs and Romani who stayed or returned after the 1999 Operation now live in what has been termed "a new apartheid."[70] The third method has facilitated the exercise of peace by utilizing a piece of territory in Kosovo for purposes of creating a US military base, Camp Bondsteel, described as "a smaller version of Guantanamo."[71] Mixing torture, surveillance and legal impunity, Camp Bondsteel is a peculiar way of implementing humanity and peace. That the Balkans is already on the edge of nowhere, and a zone of monstrosity, only serves to facilitate the provision of what have historically been 'invisible zones' of torture camps such as Guantanamo Bay in Cuba. The outsourcing of torture to zones that are missing from the 'map of civility' not only accommodates the act of torture, but also extends the possibility for continuation of the 1999 Operation and its administrative method of detention, that is, making violence invisible. In such zones, all sense of responsibility is absolved.

That law is a malleable safeguard of rights in a military intervention is also evident in the way this Operation has been implemented through the legal instrument of so-called silence procedure. The silence procedure is used when a military decision requires a consensus. It binds NATO alliance members to follow implicit consensus in actions predominantly driven by the US and the UK, thus making it easier to overcome any veto in the UN Security Council. Reaching consensus is further made easier from the perspective that members do not need to vote if in favour of the decision, but only if there is an explicit objection to it.[72] Even though sovereignty has largely lost its political importance – in that, first, the "sovereign is he who decides on the exception" and, second, the exception is dependent on the articulation of the enemy – law can be suspended and reinstated if invoked for purposes of security, such as during military interventions and legal procedures.[73] The silence procedure accommodates such political landscape by preventing debate and removing resistance as well as contested interpretations since NATO's intervention was justified in the cause of security, humanitarianism and democratic values.

Seemingly, the silence procedure is no longer reminiscent of Weizman's likening of a detective with the criminal at tribunals and courts.[74] What is embedded within the silence procedure is non-registration of any crime that may be perpetrated, since the script from the start has only one mode of speaking and interpreting – that of implicit consensus through silence. An example of this was also seen in NATO's Rambouillet Agreement, where the conditions stipulated that NATO personnel "shall be immune from all legal process, whether civil, administrative, or criminal" and that any "NATO personnel erroneously arrested or detained shall immediately be turned over to NATO authorities."[75] However, NATO's rights were inclusive of arresting individuals if considered in violation of the NATO agenda.[76] The legal agenda of the rejected Agreement, like that of the implemented silence procedure, is that the possibility that a military operation might be called a crime is eliminated from the start. Both the act of violence and knowledge about it can be constructed, edited or erased to suit political exigencies.

Law was an important element even during NATO's 'humanitarian' Operation. Procedures eliminated all international legal rights because NATO considered this territory to be a potential threat. The loose definition of 'potential threat' permits the law to be stretched to suit NATO's purposes. The elasticity of international law is evident even in the wake of NATO's 1999 targeting. The Hague Tribunal, that is, the International Criminal Tribunal for Yugoslavia – ICTY, had set up an office in Belgrade with the objective of modifying the legal statutes of Yugoslavia as well as positioning the ICTY above Serbia's national judiciary. Unlike Derrida's evocation of the Tower of Babel,[77] which allows a continual unfolding of space and place with a diversity of narratives, as well as suggesting that neither the origin nor the core of the Tower of Babel can be reached since the structure exhibits the inability to complete, translate and totalize, NATO's role seems to that of playing God, decreeing who will be the translator of truth – when it comes to both meaning and structure. The intervention facilitated through the implementation of the silence procedure, and subsequently the ICTY, is a Tower of Babel that does not allow interpretations that fail to match the NATO's motives.

What remains constant in this procedure is that its legitimization and erasure do have an implied structure of transformation, made possible by military and law. The transformation of values through the law was seen in NATO being able to differentiate at what point the intervention was enacted on behalf of 'humanitarianism' and 'security' (always endorsed), and at what point the right to intervene would stray into military acts of violence (never endorsed). The silence procedure and the legal interpretations of violence it facilitates are possible precisely because law is perceived as a contained and set structure. The US Ambassador Richard C. Holbrooke's speech during the issuing of the activation orders for the Kosovo air campaign made evident that the law is interpreted and created, by noting that "[w]e realize we are pushing the envelope and making up history as we go along."[78] In other words, practice and interpretation of law can also go beyond the written law.

Utilizing Jacques Derrida's thinking, the very belief in law, including that one may lay claim to absolute knowledge of its content, is impossible; the 'original' is always a translation.[79] To claim that NATO's intervention was humanitarian, just and legal is a matter only of perception, prejudice and translation since, following Derrida's thinking, translation is a given in language and as such the formation of meaning is contingent on alteration of that meaning.[80] The very engagement with the question, critique and meaning of violence automatically generates its mutation. Likewise, the very belief that there is a universal comprehension and understanding of violence, but that it can be exercised by only a select few, makes violence not just immanent but also permissible.

The humanitarian intervention deployed through the legal mechanism of the silence procedure was endorsed so widely despite or because of the fact that it was so inconsistently consistent. It was able to secure this endorsement thanks to the projected rhetoric that violence can be eliminated altogether as well as by pursuing an agenda that was "military without being military, political without being political, partisan without being partisan. It was all these things precisely insofar as it was humanitarian."[81] The legitimacy of the intervention, while supposedly straightforward and transparent, is inconsistently obscure. While the act of violence is made possible through its "humanitarian" agenda, the ability of that agenda to erase effects such as the Djakovica road example starts to problematize the positing of violence in the first place. The urgency behind humanitarianism thus eliminates the possibility to contest and think outside inconsistently set parameters. In other words, it is not about justice but about the immanence of law as a dialectic structure that operates precisely due to its interpretive agenda.

What makes the humanitarian rhetoric in this case even more terrifying is that aid agencies with a track record of impartiality deferred to NATO's agenda; often, they dismissed Serbian and Romani civilian casualties as not requiring their assistance.[82] The most chilling example being that in April 1999, in the midst of NATO's Operation, the historically impartial UNHCR asked assistance from NATO in order for NATO to revoke Macedonia's position that permitting the entry of around 65,000 refugees will be accommodated only if it received relief-aid from other states.[83] The relief came through NATO by building camps for the refugees. The collaboration between humanitarian agencies and military organizations is relatively recent, first appearing following the end of the Cold War. NATO is certainly turning into a single organization show, from enacting humanitarian operations to instating peace, enforcing security and assisting relief agencies while also building both refugees camps and military prisons.

The coming together of peace, prisons and justice is an example par excellence of Foucault's disciplinary societies enfolded in Deleuze's control societies. Unlike Foucault's disciplinary spaces where power is primarily applied in the present tense as a measure to predict patterns of individual behaviour, Deleuze's notion of control is more complex and subversive, constantly redrawing control in ever more encompassing webs that are both present- and future-oriented.[84] Values such as justice and freedom have been taken for granted, and their framework and social implementation left unquestioned precisely because control societies are fluid modulations accommodating of continuous and extreme reversals; the conditioning of which is comparable to a "self-deforming cast that will continuously change from one moment to the other, or like a sieve whose mesh varies from one point to another."[85]

The former US General Charles Krulak's concept of Three Block War, which arose from the deployments in Somalia, Haiti and Bosnia – meaning that US forces would end up fighting, peacekeeping and managing humanitarian tasks – not only has turned into a reality, but also has been surpassed.[86] The current disciplining is a permanent coercion, where the military is the new state prison, police, human rights agency and pseudo-forensic team. NATO's forensic capabilities were seen in its attitude towards verifying the extent of genocide and ethnic cleansing in Kosovo. The atrocities in the lead-up to NATO's 1999 Operation, including the widely reported Racak massacre executed by Serbian paramilitary forces (which set the Operation in motion) were narrated as another holocaust.[87] Deaths and expulsion did occur, but not in the way narrated by NATO. A year after the intervention, and during the ICTY, it was confirmed by Carla Del Ponte – the chief prosecutor for the tribunal – that the final death count in Kosovo was 2,788. The number was inclusive and largely composed of military fighters on both sides – the KLA and Serbian paramilitaries.

Memorials, victory and victimization

NATO's termination of the operation was marked in an odd way in Belgrade, by spectacularizing history through the erection of a memorial in the Park of Friendship on 12 June 2000. The memorial Eternal Flame (*Vjecna Vatra*) was designed by architects Marko Stevanovic and Miodrag Cvijic and sculpted by brothers Svetozar and Svetomir Radovic. The name, ironically, was borrowed from the 1946 memorial in Sarajevo. The initiative for the construction, intended to honour the defence of Serbia and ultimate 'victory' over NATO in 1999, came from the Milosevic government. The memorial was planned to stand 78 metres tall, a symbolic number recalling the 78 days NATO bombed Serbia. The project was designed in two days and the structure built in another ten days by 160 workers. At its completion, its height was significantly reduced to 27 metres.

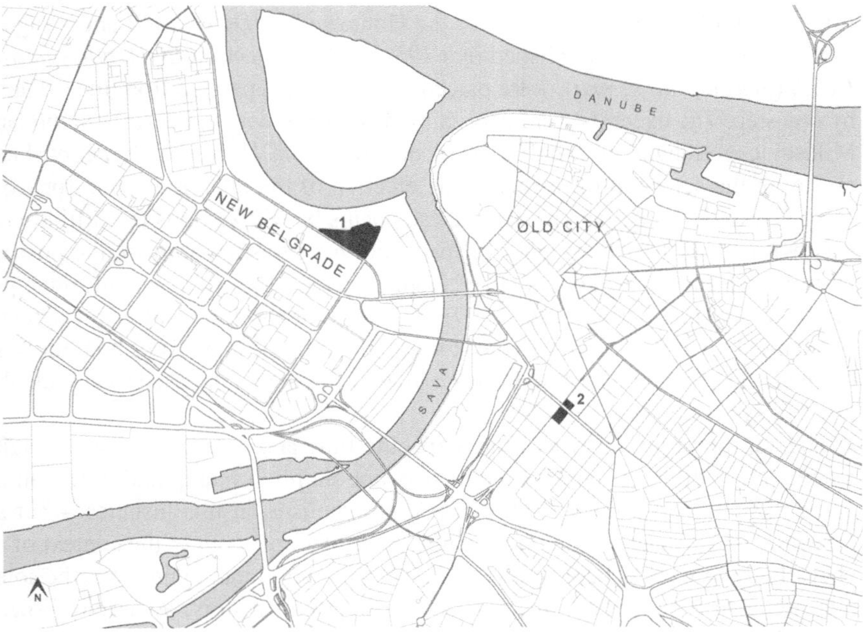

Figure 4.1 Belgrade and New Belgrade: 1 – Park of Friendship; 2 – The Generalstab complex (the Military Headquarters and the Ministry of Defence).

However, the placement of the memorial in the Park of Friendship is even more ironic. The park was the first of four large purpose–designed parks in New Belgrade,[88] located along Belgrade's foreshore, commemorating the First Summit of the Non-Aligned Movement in 1961. Its location at the confluence of the rivers Danube and Sava occupies a 12.5-hectare area of New Belgrade's total 4,100 hectares. Its large size was made possible largely through the fact that the land was owned by the state. The Eternal Flame occupies a significant place in the park, as it stands at the end of the 180-metre-long pathway called The Colonnade of Peace (*Aleja Mira*) lined with a 'living memorial' of 26 trees.[89] They were planted by the 25 heads of state who attended the First Summit.[90] The first seedling was planted by the Park's initiators – Young Foresters (*Mladi Gorani*). The London plane trees[91] (*platanus* × *acerifolia*) were spaced 8 metres apart, with a plaque placed at the base of each tree identifying the country and head of state who planted it. A sculpture by Lidija Misic was placed in the middle of the Colonnade. The indefinite form of the sculpture could allude to a tree, a flower or a bird and was intended to signify the beauty brought by peace, friendship and co-existence. In 1965, a public competition was launched to redesign the Park. It was won by the architect Milan Palisaski, with his triangular landscape design, although the entirety of this proposal was never fully realized.

The Eternal Flame memorial, which is composed of three key elements: white painted concrete base and column and the bronze flame atop, found at the end of the Colonnade of Peace, certainly dominates the surrounds due to its height as well as the odd proportions of each element. Two marble plaques with excerpts from Branko Miljkovic's poems "To the Homeland" ("*Domovini*") and "Yugoslavia" ("*Jugoslavija*"), published in a 1959 poetry collection *By Death Against Death* (*Smrcu protiv smrti*), were placed on the base, separated from the ground by one step. The excerpts were chosen by Mira Markovic, the wife of Slobodan Milosevic, notionally on behalf of the people of Serbia. However, *the people* had no say in the selection of poems or the erection of the memorial. The irony of Mira Markovic using excerpts from these poems by Miljkovic is that Miljkovic was not a supporter of any political party during Titoism. More so, the poems were written and derived out of "that" Homeland (SFRY/Serbia) and then manipulated to suit *this* Serbia (after 1999).[92] The use of the fragment "And if they killed me, I love you," from "To the Homeland" to allude to NATO's bombings is problematic, since the verse indicates a love of life that is opposed to any ideology and system.

The appropriation of ideas from a specific time and political context to suit a completely different political context is not simply a manipulation to suit a particular political project – Milosevic's nationalism, in this instance – but a revealing of the danger of taking a single verse of a poem out of the context of a poet's body of work. There is also a risk when an individual in power is the sole interpreter of a particular situation, and indirectly of history, as heroic. When that individual adopts such a role, and society participates in the historical spectacle in silence, it is an example of Nietzsche's "shrunken, almost ludicrous species, a herd animal, something full of good will, sickly and mediocre."[93]

Figure 4.2 Park of Friendship: Lidija Misic's sculpture with the Colonnade of Peace in the foreground and the 2000 Eternal Flame memorial in the background. The right-hand side visual shows the Eternal Flame memorial in 2010 with one of the graffiti reading 'Kosovo is in the heart of Serbia;' the NATO-targeted and now reconstructed Usce Tower is in the background.

Ironically, in the first two lines of the poem "Yugoslavia" chosen by Markovic, Miljkovic wrote "[a]ll that does not have fire within itself burns/ what burns becomes night/ what does not burn gives birth to day." The birth of a new day is not the sort of spectacularization of history evident in Mira Markovic's inaugural speech, where she proclaimed:

> Let this flame burn eternally as a reminder of the war that was waged by NATO's 19 countries – USA, Canada, UK, France, Germany, Netherlands, Italy, Greece, Turkey, Denmark, Belgium, Spain, Portugal, Iceland, Norway, Luxembourg, Poland, Hungary and the Czech Republic – against Serbia between 24 March and 10 June 1999. Let it burn eternally as a memory of the heroism of Serbia's defence in which all the citizens participated. Let it burn eternally and for the whole world [. . .].[94]

The Eternal Flame memorial was erected to attempt to obscure the historical fact that Serbia did not win the war against NATO. While the intent of the memorial was to feature a constantly lit flame, perhaps invoking the one in the Eternal Flame memorial in Sarajevo, nearby gas storage could supply enough power only for the opening ceremony.

The artificial gas flame and the symbolic bronze flame further showcase confusion of concept; the memorial is certainly far removed from being a marker of victory and the new age. Instead, it may be said that the memorial in Belgrade is a poor copy of Socialist Realism,[95] or that it is a second-rate architectural appropriation of the memorial in Sarajevo as well as an inflated simulacrum of history that never happened; unlike the 1946 example, when Nazi Germany was truly defeated. To draw any parallels between this memorial and the 1946 Eternal Flame memorial in Sarajevo not only short-circuits history for purposes of inflating heroism (actually victimization), but also coercively blurs the line between fact and fiction, and, in turn, helps to manage a particular understanding of events, memory and history. The history of post-WWII Yugoslavia as an example of fascism overthrown is thus conflated with the aftermath of 1999, an example of a *softer* and more purified fascism in operation.

With the memorial being erected so soon after the NATO targeting, little attention was given to placing that targeting of Belgrade in the larger context of events that occurred during the 1990s dissolution of the SFRY, or to the impact that the 1999 targeting might have on the course of history. When events occur and are not placed within the larger historical perspective or directed to critically analysing the past by projecting alternatives into the future, history remains closed. Short-circuiting history in the wake of the 1999 incursion is associated with identity, memory and history being managed through spectacularization. Mira Markovic's attempt to turn the trauma and terror associated with the NATO targeting as well as the violence generated between Serbs and Albanians in Kosovo prior to the intervention into a spectacle of nationalism not only helps flatten the understanding of history, but also exposes the danger when spectacle infiltrates politics. If history is political, then it is not a matter of safeguarding and ironing a particular rhetoric but of challenging the hegemonic notions of that rhetoric.

Architecture as partnership of peace: balkanization in vacuum

Where the irony of memorializing resilience, victory and autonomy steals the limelight is with the reconstruction of Belgrade's Military Headquarters (Building A) and the Ministry of Defence (Building B), the Generalstab complex. Despite the complex becoming an informal memorial and cultural artefact associated with Operation Allied Force, more recently the role of the complex in its past, present and future has been oriented towards eliminating any connotations that deviate from the politics of somewhat corrupt control found within the Serbian Government of the Property Directorate and the Serbian Army. Previous identifications[96] of the Generalstab complex have been concealed through a process of erosion and demolition, both of which are seemingly contingent on negation and forgetting.

The exterior of the complex was placed under a historical preservation order in 2005, as it was considered a canonical example of modernist Yugoslav architecture.[97] More recently, the Association of Serbian Architects has put forward an initiative for the structure to be listed on the UNESCO list of cultural heritage. This is in light of attempts to remove the preservation order with the emergence of an opportunity to sell the land to an international buyer for conversion into a hospitality venue. It seems that "the economic benefits surpass all others, and in the light of the constant struggle for financial stability in the national budget they [governmental institutions] see demolition and new construction as a valid reason."[98] To do this, the complex has been carved up incrementally since 2010, in the name of "structural stability and safety," as the prevalent rhetoric has it.

In this clean-up process, traces of 1999 violence are also being removed. The removal has significantly sped up in the past few years; the entry pavilion of the Ministry of Defence (Building B) was entirely removed in 2015 and the central core of the Military Headquarters (Building A) hollowed out in 2016. In order to mark this process, the front façade of the Ministry of Defence (Building B) is currently veiled in the Serbian Armed Forces' billboard, on which a quote by the late 19th–early 20th Century Serbian Field Marshal Zivojin Misic – who is seen as the most effective Serbian commander, a participant in all of Serbia's wars from 1876 to 1916 – reads, "Who can, may. Who knows no fear, advances forward." Misic's times, whilst no doubt violent, were ones of struggle to forge an autonomous way forward. The quote itself is from a time associated with distancing from Turkish rule and the overthrow of various foreign strongholds. Today, it has been deployed in a very different political context; the period of *then* to suit Serbia of *now* is a manipulation of ideas used to drive a political project in gridlock. This cunning play is further apparent with the company – Yugoimport – that sponsored the billboard. It is not because this company imports and exports defence-related equipment, but because the company carries the name of a country that no longer exists; Yugoimport was formed in 1949 when Yugoslavia was a socialist country, and one that was by the 1960s associated with NAM. Now, Yugoimport is supplying the Ministry of Defence of the Republic of Serbia with imported complex weapon systems, whilst Serbia is in the process of joining forces with NATO.

The once nationally condemned NATO attack is metamorphosing into a drawcard to join the operational forces of this pact. The deception orchestrated by politicians and the media is apparent on the façade of the Ministry of Defence (Building B). In the context of the contracts entered into between NATO and the Serbian Army Forces, the message on the banner which covers the façade of Building B is not about autonomy or implementing a distinct vision, but about manipulation and a deceit upon the Serbian population, since up to 75 per cent of them are against Serbia joining NATO.[99] The "vote against" is largely driven by the persistent memory of the effects of NATO's targeting. The bottom line is that the initiative is undemocratic in light of the majority opinion poll. While one may argue that Serbia has not actually joined NATO yet, it is an undeniable fact that the provisions made in the past 10–15 years have opened the doors for an orchestration of policies and structures deemed necessary by NATO. This is seen with Serbia's cooperation with NATO, which is most likely an indicator of Serbia losing the political right to pursue various national decision-making processes to do with the country and international relations.

While the encroachment of NATO's prerogatives has been apparent since 1999, their physical spatialization was clearly marked in 2006. That year

Figure 4.3 The Military Headquarters (Building A) in 2014 (top) and 2017 (bottom).

NATO's Military Liaison Office was opened in Belgrade, and Serbia joined the NATO's "Partnership for Peace." The pretext behind this move was for NATO "to provide advice and assistance to the Serbian authorities on reform and modernization of Serbia's armed forces, and to build a modern, affordable, and democratically controlled defense structure."[100] There is a strange placement

Figure 4.4 The billboard on the façade of the Ministry of Defence (Building B) reads, "Who can, may. Who knows no fear, advances forward."

of words here: democracy, control and defence. Stringent procedures have been implemented to do with security and defence reform, security cooperation and public information. It will soon require a magnifying glass to detect the differences between the nationalistic regime bolstered by control and manipulation of information during the Slobodan Milosevic era and the current democratically controlled government spearheaded by Aleksandar Vucic as the Serbian President that ushers in the 'Partnership for Peace.' It seems that what Virilio foresaw is coming sooner than envisaged; when addressing questions to do with the instantaneous delivery of destructive power, he responded that "if we don't watch out" it will arrive "without a real war ever having started."[101] The Generalstab complex is indicative that more recent political dispositions facilitate a time of perpetual present, and balkanization of policies and defence structures are used as means to perpetuate this vacuum.

Notes

1 Ian Buchanan, "Treatise on Militarism," in *Deleuze and the Contemporary World*, ed. Ian Buchanan and Adrian Parr (Edinburgh: Edinburgh University Press, 2006), 35.

2 Nicholas J. Wheeler, *Saving Strangers: Humanitarian Intervention in International Society* (Oxford: Oxford University Press, 2000), 11.
3 NATO, *The Situation in and Around Kosovo* – Statement Issued at the Extraordinary Ministerial Meeting of the North Atlantic Council held at NATO Headquarters, Brussels, on 12 April 1999.
4 Quoted in Leon Wieseltier, "Force Without Force," *New Republic*, 26 April 1999.
5 NATO, Press Statement by Dr. Javier Solana, Secretary General of NATO, 23 March 1999.
6 David Morrison, "Kosovo: Lies, Half-Truths & Omissions," *Labour & Trade Union Review* (November 2000).
7 Didier Fassin, *Humanitarian Reason: A Moral History of the Present Times* (Berkeley: University of California Press, 2011), 189.
8 See: Sean Murphy, *Humanitarian Intervention: The United Nations in an Evolving World Order* (Philadelphia: University of Pennsylvania Press, 1996); Sean D. Murphy, Won Kidane and Thomas R. Snider, *Litigating War: Mass Civil Injury and the Eritrea-Ethiopia Claims Commission* (Oxford: Oxford University Press, 2013); Gregory H. Fox and Brad R. Roth, eds., *Democratic Governance and International Law* (Cambridge: Cambridge University Press, 2000).
9 See: Alexis de Tocqueville, *Democracy in America, Vol. 1* (London; Vintage, 1990); Alexis de Tocqueville, *Democracy in America, Vol. 2* (London: Vintage, 1990).
10 Misha Glenny, "Only in the Balkans," *London Review of Books* 21:9 (1999).
11 Susan Woodward, *Examination of Witness FAC Report Vol. II*, 106–107.
12 Cited in "Nato Missile Video 'No Distortion,'" *BBC News*, Friday, 7 January 2000, <http://news.bbc.co.uk/1/hi/world/europe/594800.stm>.
13 BBC News, "Nato's Missile Video 'No Distortion,'" 7 January 2000.
14 BBC News, "Nato's Missile Video 'No Distortion.'"
15 Cited in James Bovard, "Kosovo Déjà vu," *The Future of Freedom Foundation*, 1 July 2003.
16 Vivian Martin, "Chapter 25: Civilian Casualties of NATO's War on Yugoslavia," *International Action Center*, 99/00, <https://iacenter.org/warcrime/25_civil.htm>.
17 Bovard, "Kosovo Déjà vu."
18 Shea notes that the video was shown at double the speed, while other sources indicate that the speed was tripled. See: Bovard, "Kosovo Déjà vu."
19 *BBC News*, "Nato's Missile Video 'No Distortion.'"
20 Paul Virilio, *Bunker Archaeology* (New York: Princeton Architectural Press, 1994), 60.
21 Cited in "Gladstone's Ghost," *The Economist*, 27 May 1999.
22 Andrew Herscher, *Violence Taking Place: The Architecture of the Kosovo Conflict* (Stanford: Stanford University Press, 2010), 108.
23 William M. Arkin, "Smart Bombs, Dumb Targeting?" *The Bulletin of the Atomic Scientists* 56:3 (2000): 49.
24 Eyal Weizman, *The Least of All Possible Evils: Humanitarian Violence from Arendt to Gaza* (London: Verso, 2011), 10–14.
25 Eyal Weizman, *Forensic Architecture: Notes from Fields and Forums/Forensische Architectur. Notizen von Felden und Foren*, dOCUMENTA (13) (Ostfildern, Germany: Hatje Cantz, 2012), 14–15.
26 Marcus A. Doel, *Geographies of Violence* (New York: Sage, 2017), 19. Also see: Gilles Deleuze, "Postscript on Control Societies," in *Negotiations*, trans. Martin Joughin (New York: Columbia University Press, 1995), 177–181.
27 Judith Butler, *Frames of War: When Is Life Grievable?* (New York: Verso, 2016), 25.
28 The federal state comprised the republics of Serbia and Montenegro from the former SFRY, which existed from 1993 to 2003.
29 Noam Chomsky, *The New Military Humanism* (London: Pluto Press, 1999), 3.

30 Noam Chomsky, *Failed States* (London: Penguin, 2017), 252.

31 Samuel Weber, *Targets of Opportunity: On the Militarization of Thinking* (New York: Fordham University Press, 2005), 9.

32 Elaine Scarry, *The Body in Pain: The Making and Unmaking of the World* (New York: Oxford University Press, 1985).

33 Jean Baudrillard, *The Gulf War Did Not Take Place* (Sydney: Power Publications, 1995). Also see: Aaron Schwabach, "Kosovo: Virtual War and International Law," *Law and Literature* 15:1 (2003): 5.

34 The MIME network is theorized by Paul Virilio and James Der Derian. It is a type of a war machine that not only assists in manipulating the line between fact and fiction so that the complexity of war and violence is reduced but also can through the conflation of the media and technology help disguise possible acts of crime. It is operationalized in the name of "security and peace," "civil values" and "humanitarianism." For Paul Virilio, the media and technology have been used to accelerate and extend the limits of conventional war strategies, where the modern world of militarization is a world in motion. For James Der Derian, it is a virtual war machine, where war fought from a distance eliminates guilt and responsibility. For more information, see: Paul Virilio, *War and Cinema: The Logistics of Perception* (London: Verso, 1989); Paul Virilio, *Speed and Politics: An Essay on Dromology* (Los Angeles, CA: Semiotext(e), 2006); James Der Derian, *Virtuous War: Mapping the Military-Industrial-Media-Entertainment Network* (London: Routledge, 2009).

35 Sultan Barakat, "City War Zones," *Urban Age* (Spring 1998): 11.

36 Baudrillard, *The Gulf War Did Not Take Place*.

37 Albert A. Nofi, *Recent Trends in Thinking About Warfare* (Virginia: CNA Corporation, 2006), 12.

38 P.W. Singer, *Wired for War* (New York: Penguin Press, 2009), 34.

39 Goran Gocic, "Symbolic Warfare: Nato Versus the Serbian Media," in *Degraded Capability: The Media and the Kosovo Crisis*, ed. Philip Hammond and Edward S. Herman (London: Pluto Press, 2000), 88.

40 Gocic, "Symbolic Warfare," 88.

41 Gocic, "Symbolic Warfare," 88.

42 Philip Hammond, *Framing Post-Cold War Conflicts* (Manchester: Manchester University Press, 2011), 122.

43 Michael Mandel, "Illegal Wars, Collateral Damage, and International Criminal Law," in *Yugoslavia Unraveled: Sovereignty, Self-Determination, Intervention*, ed. Raju G. S. Thomas (New York: Lexington Books, 2003), 308. Note: Mandel was citing the citation of this comment by Amnesty International in a report.

44 NATO/FEDERAL REPUBLIC OF YUGOSLAVIA "COLLATERAL DAMAGE" OR UNLAWFUL KILLINGS? Violations of the Laws of War by NATO during Operation Allied Force, Amnesty International June 2000, 24.

45 Michel Foucault, *Security, Territory, Population: Lectures at the Collège de France 1977–1978*, ed. Michel Senellart (New York: Palgrave Macmillan, 2007).

46 NATO – North Atlantic Treaty Organization, "NATO's Relations with Serbia," 11:59 (9 April 2015).

47 Jean Baudrillard, *The Gulf War Did Not Take Place* (Sydney: Power Publications, 1995).

48 Marjola Xhunga, "Integration in Albania: No Longer a Pipe Dream," UNHCR, 14 April 2003, <https://www.unhcr.org/news/latest/2003/4/3e9ab9a84/integration-albania-longer-pipe-dream.html>.

49 Astri Suhrke, Michael Barutciski, Peta Sandison and Rick Garlock, "The Kosovo Refugee Crisis," United Nations High Commissioner for Refugees Evaluation and Policy Analysis Unit (Geneva, Switzerland, 2000), <https://www.unhcr.org/3ba0bbeb4.pdf>.

50 Kristina Riegert, *The Image War: NATO's Battle for Kosovo in the British Media* (Örebro: Örebro University, 2003)
51 Radmila Popović, "Enemy Construction in the Media During the Kosovo Conflict: A Comparative View," *Graduate Student Workshop, Kokkalis Program on Southeastern and East Central Europe* (Harvard University: John F. Kennedy School of Government, 2001), 11.
52 Michel Foucault, *Discipline and Punish: The Birth of the Prison* (New York: Pantheon Books, 1977), 44.
53 Popović, "Enemy Construction in the Media," 11.
54 ICG, "Milosevic's Aims in War and Diplomacy," Balkans Report No. 65, 11 May 1999, 6–7.
55 ICG, "Milosevic's Aims in War and Diplomacy," 6–7.
56 Herscher, *Violence Taking Place*, 77.
57 Herscher, *Violence Taking Place*, 77.
58 Weizman, *The Least of All Possible Evils*, 99–136.
59 Herscher, *Violence Taking Place*, 111.
60 See: Esperanza Bielsa and Christopher W. Hughes, eds., *Globalization, Political Violence and Translation* (London: Palgrave Macmillan, 2009); Brian McNair, *News and Journalism in the UK* (London: Routledge, 1999); Philip M. Taylor, *Global Communications, International Affairs and the Media Since 1945* (London: Routledge, 1997).
61 Bielsa and Hughes, *Globalization, Political Violence and Translation.*
62 Butler, *Frames of War: When Is Life Grievable?* 25.
63 Paul Virilio in conversation with John Armitage, "The Kosovo War Took Place in Orbital Space," *ctheory.net*, 18 October 2000.
64 Esperanca Bjelsa, "Globalization, Political Violence and Translation: An Introduction," in *Globalization, Political Violence and Translation*, 12.
Keith Tester, "The Terminal Paradox of Globalization," in *Globalization, Political Violence and Translation*, 30.
65 Wim Wenders and Mary Zournazi, *Inventing Peace: A Dialogue of Perception* (London: I.B. Tauris, 2013), 5.
66 Der Derian, *Virtuous War*, 50.
67 Der Derian, *Virtuous War*, 50.
68 Chomsky, *The New Military Humanism*, 31.
69 ReliefWeb, *UNEP/UNCHS (Habitat) Balkans Task Force (BTF) Situation Report No. 16*, 5 August 1999.
70 Adam Wolfson, "How to Think About Humanitarian War," *Commentary* 108:7 (July/August 2000): 44.
71 Hans-Jürgen Schlamp (interviewer), "Clandestine Camps in Europe: 'Everyone Knew What Was Going on in Bondsteel,'" *Der Spiegel*, 5 December 2005, <www.spiegel.de/international/clandestine-camps-in-europe-everyone-knew-what-was-going-on-in-bondsteel-a-388556.html>.
72 Paul Gallis, "NATO's Decision-Making Procedure," *CRS Report for Congress*, updated 5 May 2003.
73 Carl Schmitt, *Political Theology* (Chicago: University of Chicago Press, 2005), 5. Also see: Giorgio Agamben, *Homo Sacer: Sovereign Power and Bare Life* (Stanford: Stanford University Press, 2008) and Giorgio Agamben, *State of Exception* (Chicago and London: The University of Chicago Press, 2005).
74 Weizman, *The Least of All Possible Evils*, 99–136.
75 United States Institute of Peace, "Appendix B: Status of Multi-National Military Implementation Force – 6. (a) and 7," *Interim Agreement for Peace and Self-Government in Kosovo Rambouillet*, France, 23 February 23 1999, <https://www.usip.org/sites/default/files/file/resources/collections/peace_agreements/kosovo_ramb.pdf>.

76 United States Institute of Peace, *Interim Agreement for Peace and Self-Government in Kosovo Rambouillet*.
77 The structure, according to the Bible's Book of Genesis, was an enormous tower built in the city of Babylon. The citizens of Babylon, who spoke a single language, decided their city should have a tower that would reach the heavens. The intent of this structure was not to glorify God, but to show the strength and skills of the people. The Book of Genesis then relates how God, displeased with the builders' intent, came down and confused their languages.
78 Cited in Michael J. Glennon, *Limits of Law, Prerogatives of Power: Interventionism After Kosovo* (New York: Palgrave Macmillan, 2001), 32.
79 Jacques Derrida, "Des Tours de Babel," in *Difference in Translation*, ed. Joseph E. Graham (Durham: Duke University Press, 1985), 165–207.
80 Derrida, "Des Tours de Babel."
81 Clifford Orwin, "Humanitarian Military Intervention: Wars for the End of History?" *Social Philosophy & Policy* 23:1 (January 2006): 200.
82 Tony Vaux, *The Selfish Altruist* (London: Earthscan Publishing, 2001).
83 Michael Barnett, *Empire of Humanity* (New York: Cornell University Press, 2013), 188.
84 See: Foucault, "Of Other Spaces (1967)," 13–29; Deleuze, "Postscript on Control Societies," 177–181.
85 Deleuze, "Postscript on Control Societies," 178–179.
86 Frank G. Hoffman, "Complex Irregular Warfare: The Next Revolution in Military Affairs," *Orbis* 50:3 (Summer 2006): 398.
87 Olga Mitrovic, "Is the Use of Force Ever Legitimate in Order to Pursue Humanitarianism?" *Western Balkans Security Observer* 5:18 (July–September 2010): 84.
88 The other three parks were the Park of Art, the Sava Park and the Entertainment Park. Greenscapes were integral to New Belgrade's urban plan. However, these zones often remained open grassed fields due to lack of funds to design them.
89 Ivana Nesković, "Park Prijateljstva u Novom Beogradu," *Nasleđe* 12 (2011): 203–216.
90 The 25 states were Afghanistan, Algeria, Burma, Cambodia, Ceylon, Congo, Cuba, Cyprus, Ethiopia, Ghana, Guinea, India, Indonesia, Iraq, Lebanon, Mali, Morocco, Nepal, Saudi Arabia, SFRY, Somalia, Sudan, Tunisia, the United Arab Emirates and Yemen.
91 The choice of trees also elaborated the 'third way' way theme. The London plane tree is a hybrid of Oriental Plane tree (*Platanus orientalis*) and American sycamore tree (*Platanus occidentalis*).
92 Slobodan Kostić, "Simbolika Večne Vatre," *Vreme 493*, 17 June 2000.
93 Friedrich Nietzsche, *Beyond Good and Evil: Prelude to a Philosophy of the Future* (Harmondsworth, Middlesex: Penguin, 1973), 62.
94 Biljana Vasić, "Postavljanje spomen-obeliska u Beogradu," *Vreme 493*, 17 June 2000.
95 It is an architectural style that arose in the Soviet Union and became the national style from early 1930 to late 1980s. After WWII, Socialist Realism became the official architectural style of most of the European socialist/communist countries. Yugoslavia was an exception to the rule after its parting with the Soviet Russia in 1948.
96 These identifications are inclusive of the design of the complex being inspired by the WWII Sutjeska Offensive and Henri Bergson's philosophical conceptualization of matter and memory through the diagram of a cone, to the association of the complex with NATO's targeting.
97 Dušan Miličević, "Ruševine koje svima vezu ruke," *Blic*, 30 April 2009.

98 Milošević, "The Ghosts of the Past, Present and Future: The Case of the Army Headquarters in Belgrade, Serbia," 51.
99 Ana Roland and Bojana Jelovac, "Građani: Ne u NATO . . . ali se možda mora," *Blic*, 6 April 2015.
100 NATO – North Atlantic Treaty Organization, NATO's Relations with Serbia.
101 Paul Virilio, *A Landscape of Events* (Cambridge, MA: MIT Press, 2000), 96.

5 Circulating violence

Decomposition, dispossession and control

The less visible violence

Balkanism as excess in Kosovo

Balkanism is a discourse of perceived excess where the identity of the region allows for a complete breakdown and disruption of any norms, including those which articulate a clear definition and belonging to either the West or the East. In this signifying process, Balkanism is implicated in the idea that its identity needs to be continually *abjected* in order to reinstate its borders, one that resonates with Julia Kristeva's framing of abjectness: "I expel myself, I spit myself out, I abject myself within the same motion through which 'I' claim to establish myself."[1] As an *abjected* zone, Balkanism proffers the opportunity to engage with questions of sanitation implicit in Anglo-American and Western sanitary dispossession and decomposition of infrastructure and resources deployed for purposes of 'normalizing' Serbia post-1999. Thus, Balkanism will not necessarily be used to address questions to do with normalization, which in this case is reflective of processes of regulation; more readily, the discourse will be thought as an opportunity to address "*for what?*" purposes are these regulations being implemented, and "*what is not being said?*" In terms of Kosovo, the question is what was it that instigated the intervention when other notable examples of violence around the world such as the conflict in Democratic Republic of Congo that started in 1996, or the 2 million Sudanese refugees since 2003, apparently did not merit the same urgency of intervention or media exposure? To intervene in Kosovo was, in contrast, a matter of overwhelming importance.

Historically, for both the Albanians and the Serbs, the territory of Kosovo (or the Autonomous Province of Kosovo and Metohija, as it was known in Serbia and former Yugoslavia in the 20th Century) is identified with notions of loss and suffering. For the Serbs, the Battle of Kosovo in 1389 has taken on mythical proportions, bearing the significance not only of a battle lost against Turks and the 500-year Ottoman rule that would follow, but also the loss of territory associated with the emergence and flourishing of Serbo-Byzantine architecture during the Nemanjic Dynasty (1160s–1371). The architectural style was implemented in churches and monasteries and located predominantly throughout Kosovo. The architectural heritage is synonymous with the national identity of Serbs and the seat of their ancestors. For the Albanians, the belief is that they are the original Illyrian settlers of the region, that Albanian is linguistically closest to Illyrian, and thus that the territory is originally and rightfully theirs.

Despite these respective positions, up until the turn of the 20th Century, Albanians and Serbs in Kosovo mostly collaborated in their fight against the Turks, the common enemy since the 14th Century. With the defeat and dissolution of the Ottoman Empire in the 19th Century and early 20th Century, Serbia saw Kosovo as justly hers considering the cultural and artistic heritage established in this region in the Middle Ages. For the Albanians living in Kosovo, the Serbian reoccupation was simply an instatement of a different version of colonial occupation, which would be shored up by Serbian and Yugoslav government attempts to accord prominence to the Serbian version of history.[2] Such perception, inclusive of lack of cultural and linguistic commonality between Serbs and Albanians, led to tensions even before WWII; the desire among Kosovo Albanians (now also known as Kosovars) was to assert independence and become a part of Albania.

Up until 1968, Serbs in Kosovo were given preference in all areas of decision-making, which provoked dissent and nationalist sentiment among Kosovo's Albanians. However, once the province was redefined as a sociopolitical entity under the 1974 Yugoslav Constitution, the majority Albanian population in Kosovo were given substantial rights. They held a veto power over decisions made by the Serbian parliament to do with Kosovo, unless that decision-making related to matters of significance to Serbia.[3] Having said this, Kosovo held power of veto over decisions made by the Serbian parliament on matters to do with the province. With the 1974 Yugoslav constitutional changes, Kosovo was granted near-republic status, including representation at the federal level and the power to issue its own constitution.[4] Rights to autonomy extended so far as to permit the University of Pristina to teach course content in Albanian, the only institution of higher education in the former Yugoslavia to do so.[5] The catch-22 was that despite the Kosovo Albanians' notable autonomy and ability to exercise self-rule, they did not have the same technical right to secede that other federation member states enjoyed.[6] It did not assist that the economic development within republics and autonomous provinces was very uneven, with Kosovo being one of the more undeveloped areas.

Despite economic assistance received from other more developed republics such as Croatia and Slovenia, and developmental gains within Kosovo, the significant demographic growth within the province was eroding the possibility for economic prosperity. In 1981, the population was ethnically relatively homogeneous (91 per cent Albanian) and had grown by 23 per cent since 1948, one of the highest rates of increase in Europe. In the wake of an economic recession and high unemployment rates, the 1981 student demonstrations motivated by socioeconomic disparity also spurred a wave of Albanian nationalism. The ethnic and nationalist dimension was evident in student slogans: "We are Albanians, not Yugoslavs," and "Unity with Albania."[7] There was a reversal of discrimination;

from Albanian oppression to Serb. According to *New York Times* correspondent Chris Hedges, the period from 1966 to 1989 saw about 130,000 Serbs leave the province due to harassment.[8] Discrimination was reported in workplaces and social life; people were threatened with rape and forced to sell their agricultural land below its value in order to escape.[9]

With Milosevic coming into power in mid- to late 1990s, including his adoption of the Kosovo Serb cause in 1987, a notable shift in Serbia's control of Kosovo occurred, primarily through alteration of the constitution. The alterations stipulated the transfer of all state companies into Serbian ownership as well as the withdrawal of the Albanian curriculum in schools. The early 1990s attempts by Albanians in Kosovo to gain independence were not internationally supported. The failure of the 1995 Dayton Peace Agreement to accommodate the rights of Albanians in Kosovo provided a pretext for the escalation of violence and further nationalization of the cause, including the formation of the KLA. The period from 1989 to 1998 saw violence in Kosovo perpetrated to varying degrees by both sides; these events in their totality were portrayed as critical to NATO's 1999 targeting decision.

However, according to Eric Hobsbawm, the 1999 NATO Operation was the "first war fought under conditions of [. . .] consumer sovereignty."[10] Its humanitarian agenda was brought to bear on behalf of an ethnic group, the Kosovo Albanians. On a humanitarian pretext, a calculated value judgment was made on behalf of an ethnic group rather than asserting humanitarianism on behalf of all (irrelevant of nationality) who are in need of humanitarian assistance. The air attack further created a situation for a ground intervention, made all the more possible by projecting the context using transitional and post-socialist tags in the name of Western values.

Immobilizing movements in Serbia

While the destruction of infrastructure in Belgrade was considerably smaller than in Vukovar or Sarajevo, the agenda may have been more complex in that targeting was oriented towards severely degrading transport routes and media as well as other lines of communication and industries. In effect, the focus may have been on strategically hamstringing the economy and its circulation to a point where rebuilding the system and securing the country's safety was possible only through recourse to foreign loans and security measures. Belgrade, and Serbia, were introduced to Western programmes of security and surveillance in the form of both long- and short-term operations. The long-term included the imposition of UN Security Council resolution 757, adopted in May 1992, as well as – amongst other bans – economic and air travel embargoes (unless this was for humanitarian needs) on Serbia and Montenegro during the 1990s SFRY disintegration. The restriction on mobility, and the economic hyperinflation that followed, accommodated a series of short-term operations that reached their apex during the 1999 NATO targeting.

One of the coercive aspects of NATO's intervention[11] was oriented towards disciplining and controlling a country that was not consumer capitalist. On 23 March 1999, one day before NATO's strikes began, US President Bill Clinton stated the importance of economics to the intervention:

> [if the US is] going to have a strong economic relationship that includes our ability to sell around the world, Europe has got to be a key. And if we want people to share our burdens of leadership with all the problems that will inevitably crop up, Europe needs to be our partner. Now, that's what this Kosovo thing is all about.[12]

Clinton's position was reaffirmed in NATO's Rambouillet Agreement. The stipulation of free market principles hints at the later incursion being driven by the desire to transform economic values in the direction of the market economy.[13] This is significant, as the FRY (composed of Serbia and Montenegro) was one of the last remaining countries in Europe resisting the shift towards consumer capitalism, a shift that gained momentum with the fall of the Berlin Wall followed by the disintegration of the USSR and the SFRY in the 1990s. NATO's strategy was to deploy an array of tools in the lead-up to the intervention,[14] the 1993 sanctions imposed on the FRY by the West being in part a preparation for NATO's 1999 Operation. That targeting exacerbated the 1990s economic slump from which the country was still attempting to recover as well as the instability left over from the economic stress of the 1980s. In 1993, the unemployment rate was 50 per cent.[15] In 1994, Serbia's inflation reached 313 billion per cent.[16] By 1995, 500,000 refugees from Croatia and Bosnia-Herzegovina settled in Serbia, an added contribution to the overall economic and political collapse of the country.[17]

The 1999 bombing's physical and psychological impact was extensive because the targeting was directed at not just military-oriented targets but also the everyday lives of the civilian population. This was done by targeting key infrastructure points in order to sever connections between places and immobilize the citizenry. The effect was balkanization of the country in terms of access to resources such as water and electricity. According to Julian H. Tolbert, a pilot and combatant in the operation, NATO disrupted "power to 70 percent of the population on 2 May and continued to turn out the lights through soft bombing of electrical transformer yards [in an attempt] to drive a wedge between Milosevic and the population."[18] The NATO campaign was a coordinated military matrix of "[c]overt operations, psychological warfare, information operations" in an attempt "to threaten what air war planners labelled Yugoslavia's 'centers of gravity'."[19] These "centers of gravity" attracted not only media-military-finance attacks, but also attacks directed at civilian life and all that facilitated their everyday life.

Devastation of numerous industrial facilities in Serbia took place, such as the Zastava car plant in Kragujevac and the Lola Utva aeronautical factory in Pancevo, and overall, 600,000 workers lost their jobs. The degradation of infrastructure included 41 bridges and four overpasses destroyed or heavily damaged.

The destruction of bridges also affected water mains, leaving many areas without water. NATO's use of carbon soft bombs meant that power blackouts were an everyday occurrence, further disrupting water reticulation. Overall, the intervention destroyed 100 per cent of Serbia's petroleum refining capacity, which, in turn, affected 24 fuel, oil, chemical/petrochemical industry and/or storage sites; water supply lines; rail facilities; ten agricultural complexes; and a fertilizer plant.[20] The specific damage to power-generating infrastructure affected four power stations and a coal mine, 16 substations and seven long-distance transmission lines.[21] The devastation of electrical transmitters had a domino effect on food supplies and the provision of medical care as well.

Despite US General Wesley Clark's statement that the Kosovo intervention had nothing to do with resources,[22] the reality is somewhat different. NATO's bombing of electrical systems in Kosovo in 1999 used laser-guided weapons and graphite soft bombs designed to generate massive short circuits and fires, which led to de-electrification of the cities.[23] Apart from demodernizing[24] Serbia above ground, below-ground resources were also depleted because the high-tech projectiles that were used during NATO's Operation contained depleted uranium (DU).[25] The ammunition that contained DU was PGU-14 API 30 millimetre rounds fired from Gatling guns mounted on A-10 Warthog 'Tankbuster' aircraft.[26] While DU munitions are not prohibited by international law, Article 35(3) of Additional Protocol I to the Geneva Conventions prohibits "employ[ing] methods or means of warfare which are intended, or may be expected, to cause widespread, long-term and severe damage to the natural environment."[27] This legal doublespeak is demonstrated by DU munitions not being banned by international law, but warfare that has a long-term impact on nature being prohibited.

Sanitary decomposition and limitless destruction

Even though NATO's projectiles contained DU, the strikes were globally described as "surgically precise," words that carry implications of not only health but also extreme accuracy. Considering that in this sanitized war only 29 per cent of the 28,000 bombs and missiles used were precision-guided,[28] there is a deceptive legitimization of NATO's violence. With large numbers of these cluster bombs still lying in the ground where they fell, and despite the bombing ending in June 1999, people are still being killed as undetected bombs explode.[29] In the intervention's 'sanitary' strikes, technology became an act of violence directed to destroy territorial and urban infrastructure, as well as human bodies, even after its conclusion. The effects from use of DU munitions are debilitating once the DU enters the body through inhalation and ingestion. The air as the primary substance of life has been utilized as a zone of combat.[30] The particles released into the body cause breakdown of cells, tissues and organs, destroying proteins and enzymes and damaging DNA. This kind of damage in the reproductive organs can lead to genetic hazards that can be passed on from generation to generation.

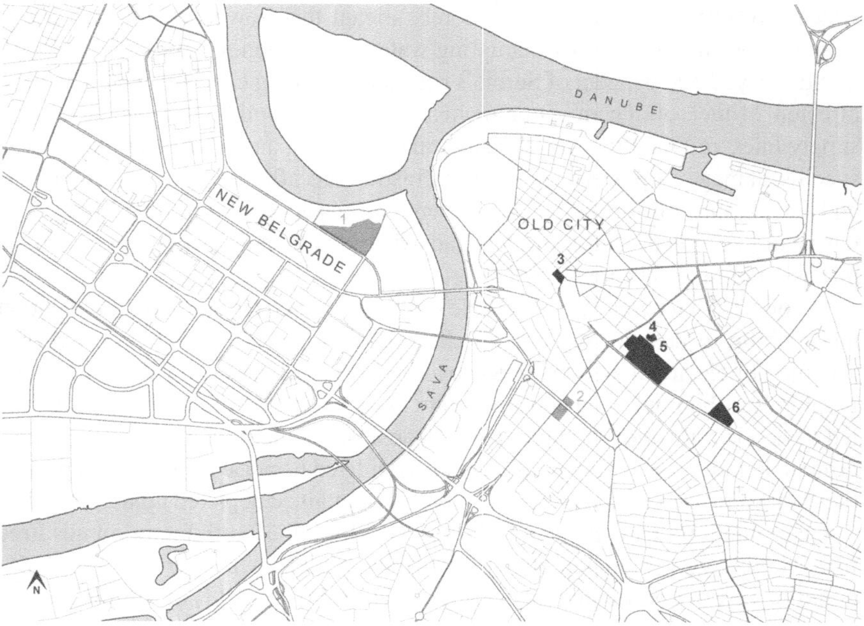

Figure 5.1 Belgrade: 3 – The Republic Square; 4 – RTS; 5 – Tasmajdan Park; 6 – Vuk's Monument train station.

Wars are no longer fought only for causes of religion or territory. Part of the process of NATO's balkanization seems to have included creating new sites of death from radiation that will remain a silent killer years after the 1999 targeting. This is no longer an example of NATO's attempt to control information as part of MIME or to dominate the air (considering that Operation Allied Force was fought from the air alone). Rather, with the current nuclear weapons, genocide becomes less detectable; the Balkanist body is eliminated permanently and silently. Its invisibility not only poses challenges to representing that violence, but also depoliticizes its pervasive effects despite their rapid spread.

The less visible death and violence brought into play by NATO's targeting is notably different from the sites of death and spatial representation via cemeteries. It is different from the 18th and 19th Century cemeteries where the focus on hygiene also conferred significance on corpses, with sites of death and cemeteries occupying an important place in modern urban infrastructure. Up until the 18th Century, cemeteries in Europe and the West occupied land close to the centre of a city or town, but thereafter they were relocated to the peripheries. In Belgrade, there has been a different siting of death and cemeteries; perhaps this may be because the city has been destroyed over 40 times in the Common Era.

Today's Republic Square (*Trg Republike*) holds traces of the 1999 mass protests of the Songs Sustained Us gatherings, including the reinterpretation of public spaces by exposing the national and international political rhetoric related to

NATO's Operation, as well as more immediate and distant histories of violence. The square is currently being reconstructed, with the proposal showing a very formalized space oriented towards eliminating the formation of crowds; it is a space to pass through rather than congregate and sit with all the previous raised planter beds, also used for seating, removed. This is significant considering that the square is a signifier of space not being fixed or contained, but that events – official and dissident – are moments of identity renewal.[31] Historically, the square has been used not only as a promenade, but also for events ranging from various celebrations to displays of fashion and art. In times of conflict and contestation, the space has been appropriated for public demonstrations. This space held particular importance during gatherings against the SFRY disintegration in the early 1990s, with the 1996–97 period experiencing escalated protests against the increasingly nationalistic Milosevic rule. During that period, the square was referred to as the Freedom Square (*Trg Slobode*).

Should the remodelling of the square follow the design proposal, the newly stone-paved and levelled space is to retain the 1882 monument to the Serbian ruler Knez (Prince) Mihailo on one end, and two rows of six trees will be planted on the other end. The monument, built in a Classical Renaissance style, fronted the 1869 National Theatre designed in the late Renaissance style, and was associated with the official break from the Ottoman Empire during the rule of Knez Mihailo. Both the theatre and the monument gave shape to the Republic Square, which at the time was known as the Theatre Square. In the vicinity of the monument, two plaques are designed to be added – to commemorate the location of the Istanbul Gate (*Stambol-kapija*) and the death of 976 Red Army[32] soldiers who died when the Nazis bombed National Square in 1944 battle. With Yugoslavia parting ways with Soviet Russia, the monument to the soldiers was removed in 1949 and the buried bodies moved to a different part of Belgrade in 1951. In 2019, there is a need to side with Russia again, and architecture is a means of doing this.

Figure 5.2 The Republic Square before the reconstruction, and now in the midst of reconstruction in 2018.

The Istanbul Gate, which was surrounded by deep trenches filled with water and tall posts stretching from the Danube to the square and looping down to the river Sava, is significant not only because it was used during the Ottoman Empire as a defensive system against foreign attack, but also because immediately behind the gate's three main openings, an area was utilized by the Ottoman Empire as a public execution ground where the living bodies of those showing open revolt against the empire were pierced on stakes. Seemingly, the commemoration of this space is associated with the displacement of one political orientation, the Eastern Ottoman, with another, the Western European and the forthcoming balkanization of the region, as well as a reminder of the violence endured during that period.

Where the gate's massive stone blocks, deep trenches and tall posts once sat facilitated space for the construction of the current day Republic Square. However, it was only at the start of the 20th Century that the square took on a particularly cosmopolitan function, with the erection of the National Museum in a Neo-Renaissance style. The Museum, along with the National Theatre, enclosed the square, giving it a public status. While planning for an urban public space that would accommodate official parades occurred during Austro-Hungarian rule,[33] the initiative was fully realized in the public gathering on 25 November 1945. The first public address of the newly formed socialist Yugoslavia was made by Tito from the terrace of the National Theatre, and the square became known as the Republic Square. While it is unknown whether the name of the square will remain

Figure 5.3 The unreconstructed section of the Radio Television of Serbia (RTS) building.

with the current reconstruction, and if so for how long, what is more certain is that the 2018–19 remaking of the square has allowed an unearthing of history considering that the area occupied by the square is archeologically rich as is in the vicinity of the site of the Roman ruins and forum. In September 2018, the excavations have allegedly led to unearthing the foundations of the Istanbul Gate, and the Roman graves. All works have been suspended until the legitimacy of these findings has been verified. Undeniably, this may not only affect the completion of the works within 420 days but also affect the proposed design. The period may also be extended should the largely unexplored underground caves, canals, tunnels and underground waters affect the construction.

The merging of sites of death and cemeteries and hygiene also occurs in the current day and centrally located Tasmajdan park in Belgrade's Old City. As well as having been hollowed out by stone quarries, Tasmajdan also has several natural subterranean caves, one of which is about 7 million years old. One of the caves was used to shelter Belgrade's population against German and Austro-Hungarian attacks during WWI.[34] Another underground layer of Tasmajdan and its surroundings was the Turkish and Austro-Hungarian cemetery (depending on the period, both empires ruled Belgrade).[35] Likewise, the area of Tasmajdan was used during the 19th Century for burial grounds, and thus contains the graves of Orthodox Christians, Catholics, Lutherans, military personnel killed during wars and those who committed suicide as well as non-Christians.[36] When the new Orthodox cemetery was built in 1886, only the remains of the affluent Orthodox adherents were transferred, while the others remained buried beneath the park.[37] The period immediately after WWII saw the burial of many German and Russian soldiers beneath the Tas stadium in Tasmajdan Park.

The area allocated for burial beneath Tasmajdan is thus a hollowed-out space of death and violence on top of which the modern day foundations of Belgrade were built. This area also contains a more recent cemetery brought into existence by NATO's 1999 targeting of the government-owned RTS building, which resulted in 16 deaths and 18 workers being injured. The RTS building remains in ruins, and the adjoining Tasmajdan Park houses a small memorial to mark the site of the deaths.

The presence of DU makes it less possible to name the source of death, commemorate it and clearly locate it since its spread has no limits. The links between DU and health disorders such as cancer spurred by NATO's use of DU are known as the Balkan syndrome. Considering that many of the NATO targets were infrastructure-based, the targeting has affected production of goods, services and overall economic budget in the short and long term as well as created environmental damage. The bombing of oil refineries destroyed 57 per cent of petroleum reserves as well as leaving a 20-kilometer-long oil slick in the Danube.[38] Air strikes on the oil refinery in Pancevo, a city only 20 kilometres northeast of Belgrade, set alight an estimated 80,000 tons of oil and oil products. This was not just a health hazard for the population, but also a threat to aquatic life in the Danube.[39] Reports show that workers at the refinery have since suffered from liver cancer due to exposure to high levels of the toxic chemical vinyl chloride monomer (VCM).[40]

Figure 5.4 A small memorial "Why?" in Tasmajdan Park to mark the deaths of those who were killed during the targeting of RTS.

In Novi Sad, a city located 90 kilometres northwest of Belgrade, the Danube was heavily contaminated after strikes hit a factory. About 73,000 tons of crude oil and oil products were reported to have leaked into the Danube there.[41] A few months after the bombing, concentrations of mercury in the air were 500 times more than the 1 mg/m3 allowed. This is notable since mercury is the third most hazardous substance listed on the US Department of Health and Human Services Agency for Toxic Substances and Disease Registry.[42] The effects of this chemical warfare were not only confined to Serbia but also impacted neighbouring countries such as Romania and Bulgaria.[43] Since Kosovo, the US and UK militaries now require their personnel to wear protective masks when operating in all the areas affected by DU.[44]

The dressing up of NATO's targeting as sanitary strikes is a different mode of disguising death to the one seen in 18th and 19th Century cities. Not only do drone attacks eliminate the potential for the intervention to be called an environmental catastrophe, being able to target from a distance accommodates the ability to control representation of both death and destruction. Such capacity facilitates the power to negate the value normally attributed to human, animal and plant life. That this is permissible is due to this region of the Balkans already being an anomaly, and thus of no representational value. The 21st Century military operations extend the current understanding of war and violence.

Apart from urban infrastructure monitored, demodernized and contaminated by graphite soft bombs and missile-delivered munitions containing DU, a more covert impact of these strategies is that they affect all forms of life. The use of DU munitions affected not only life on the ground but also all that is below the ground, by damaging deep soil health. Apart from the destruction of vegetation and contamination of the topsoil, which can take thousands of years to regenerate, the effects of DU include negative impacts on the reproductive cycles of amphibians, with potential to disturb the migratory corridors of the birds which feed on them.[45] Arguably, the acceleration of environmental destruction is yet another form of violence. While buildings might remain intact or be reconstructed, the biosphere is slowly being destroyed. NATO's Operation, according to *Guardian* journalist George Monbiot, was

> perhaps the dirtiest war the West has ever fought. NATO's scorched earth policy, which seeks to destroy Milosevic's armed capacity by destroying everything else, places the Alliance firmly on the wrong side of the Geneva Convention. For a war which targets chemical factories and oil installations, which deploys radioactive weapons in towns and cities, is a war against everyone: civilians as well as combatants, the unborn as well as the living. As such, it can never be a just one.[46]

The New World Order appears to be based on creating new hidden sites of death by destroying ecological systems and natural resources. Earth is no longer a surface to be cleared (as suggested in Monbiot's quote) or associated with "rubbleization" (as suggested in Martin Coward's understanding of urban destructions, where the built environment is reduced to rubble), but a surface to be penetrated. The deployment of DU munitions has contaminated soil, water and biosphere in the targeted territory.

The clean-up of contaminated soil would require removal of the first foot of topsoil,[47] a process which neither Serbia nor the external human rights agencies has the funds or the will to undertake. When attempts were made to remove some of the munitions, such as the cluster bombs, a number of which remain unexploded and dispersed throughout the territories of both Serbia and Kosovo,[48] the US military used local civilian teams to disarm them. Even the landmines are hard to detect due to their distribution being random; considering that they were remotely launched from US munitions system and the GPS.[49] Civilian lives were not only targeted during 1999, but now come into play again as the deadly debris of violence needs to be removed; to survive is to be exposed to a permanent possibility of death. The contamination of the environment is an ever-present concentration camp of slow death experienced in this Balkanist context.

While planned destruction through management of air and climate in the form of "atmo-terror" has been present since WWI (noted for its gas attacks),[50] current atmospheric destruction is harder to put on trial. This is particularly the case because the US, as the majority NATO member representative, cannot be legally held accountable, as it refused to sign the Ottawa Convention Agreement which, amongst other stipulations, prohibits the use of cluster bombs.[51] Another reason for NATO having the upper hand in this 'sanitary' violence is that it is often difficult to distinguish the environmental damage caused by NATO from – for example – the sedimented pollution present in the Danube and the toxicity that has been building up from the 1960s. Such indistinguishability is reflective of Adrian Lahoud's comparison of the evidence of the aftermath to the "remains of an event that are missing."[52]

The actual extent of the toxicity is unknown, yet the ongoing nature of the effects is evident in the Romani refugee camps in Kosovo. These refugees are those who have not managed to flee to Germany, Italy, Serbia, Montenegro or Macedonia; over 100,000 Romani fled in a period of three months after peace enforcement started in June 1999.[53] The remaining Romani in Kosovo are not only discriminated against by the local Kosovo Albanians,[54] but are also exposed to the effects of radiation as they have been housed in UN camps built on toxic wastelands. Those most prone to exposure are newborns, who either die from it or suffer irreversible brain and organ damage, and have a maximum life expectancy of 30 years. Despite the World Health Organization issuing a warning in 2000 to the United Nations Interim Administration Mission in Kosovo (UNMIK) to evacuate the Romani from the camps, no action was taken.[55]

New enactments of violence are harder to see, and their effects are longer term. They are implemented through chemical and environmental weapons, air- and ground-launched missiles. Operation Allied Force is synonymous with (in)visible violence as well as with the elimination of natural, territorial and bodily resources (physical and psychological). Seemingly, the less visible aspects of this New World Order, with the urban wasteland – ecocide – is that sites of death are permanently present even though their spatial borders and their visibility are eliminated. The river Danube, for example, Europe's most important west-east waterway, is in areas dead even though drinking water is sourced from the same river for about 10 million people. Here, balkanization is pursued through the ecosystem in an attempt to silently extinguish all that is still resisting complete conquest. Unlike the Cold War's nuclear strategy under which peace was maintained on the basis of deterrence, the 1999 incursion used the ecosystem as a weapon endlessly extending radiation's toxic effects. However, the flaw in the strategy is that this type of violence cannot be controlled; even if its effects are more present in the Western Balkans, the potential spread of radioactivity is global. The ecosystem has become a necessary weapon of control, as well as its destructive force.

Reintegration via dispossession

Coercion, data and militarization of the economy

NATO undertook a strategic shift between 1967 and 1991 as the US pursued changes in military policy that were needed to accommodate more flexibility; one of which was the integration of conventional forces with nuclear weapons.[56] One could speculate that NATO's 1999 Operation is an example of one of these flexibly integrated post–Cold War doctrines where the ecosystem is instrumental in the instatement of security and defence economics. However, to achieve these aims, a fertile field needed to be created for sowing belief in 'humanitarian' actions and diplomacy, and thereafter normalization and reintegration into the civilized West and Anglo-America. The economic normalization is inclusive of the intervention having a notable effect on general living standards in Serbia, with poverty levels increasing from 33 to 63 per cent immediately after the war.[57] This was more or less inevitable in light of the loss of 250,000 jobs resulting from the targeting of infrastructure, as indicated in World Bank and IMF reports.[58] Kosovo was the last remaining piece of Yugoslavia balkanized for purposes of eliminating counter economies and industries.

The targeting by NATO not only eliminated the ability of various industries to compete with Western firms, as they had done prior to the Yugoslav disintegration, but also left the recovery and rebuilding of the systems dependent on loans from the World Bank and IMF. The pace of control has picked up momentum particularly since 2011. The list of NATO acquisitions is lengthy. Notable key purchases include Serbia's Chemical, Biological, Radiological and Nuclear Training Centre in the city of Krusevac (southern Serbia), which in 2013 was turned into a Partnership Training and Education Centre. The change of ownership and use has also come with moves to make all activities within the centre transparent to NATO allies and partners. Additionally, work began on decommissioning about 2,000 tonnes of ammunitions in Serbia.[59]

When the less visible tactics of its operations are uncovered, it becomes all the harder to believe that NATO's concern was for the upholding of human rights above all else. If so, it was a twisted ideology of humanity and peacekeeping. If anything, NATO's humanitarian agenda may be seen as strategic. The intervention was oriented towards sufficiently damaging all the infrastructure systems and industries so that any reconstruction would be dependent on accepting foreign loans. In order to overcome the effects of the intervention, the process of reconstruction necessitates a form of dependence on conditional foreign bank loans and foreign control. That governing lies in the hands of economically stronger countries; the US has 17 per cent of the IMF vote, and only 15 per cent is required for a veto. Such a structure ensures control and spread of economic debt through dependence without the resort to imperialism. According to Naomi Klein, neoliberalism tends to record a surplus during times of catastrophic events, whether those be natural disasters or political ones.[60]

The ongoing reintegration is driven by economic control and privatization as well as ownership of industries and resources; it facilitates the transformation of Serbia into a balkanized enclave purely dependent upon foreign investors and on the Western military. The attack on Belgrade was an attempt both to extinguish alternative lines of communication, including the economic flows of the FRY, and to initiate a relationship of dependence upon the West. The array of infrastructural and economic interests that lay behind targeting Serbia and Belgrade's alternative lines of communication suggests that NATO's campaign is connected to a whole theatre of violence; not simply destruction and paralysis, but also a particular implementation of balkanization where systems that sustain the life of a city are balkanized just enough to coerce a surrender.

NATO's intervention has resulted in the militarization of the economy. The rebuilding of Belgrade's destroyed buildings and infrastructure is now dependent upon transforming Belgrade and Serbia according to *free market principles*, establishing consumer culture, privatization and security. Law is an important aspect in these processes of remodernization and rebalkanization. Its significance was noted in 2005 when ICTY prosecutor Carla Del Ponte said that while "the ICTY and other UN organizations are not profit-making bodies, [. . .] they, and the ICTY specifically, facilitate profit-making for others."[61] In other words, she was not associating the military intervention and the legal tribunal with morality and justice, but rather with creating a fertile field for investments.[62]

The reconstruction of Belgrade after 1999 has been oriented towards an inequitable incorporation of its territory and urban infrastructure into the wider European and global network for the purposes of Western defence and security. However, before that incorporation could be completed, and in order to deal with the rogue Balkan country, the military intervention drew on 59 airbases in 12 countries, 941 fixed-wing aircrafts and 279 helicopters.[63] Seemingly, part of the purpose of the Kosovo intervention went beyond expanding the territory available to be used when required by NATO and extended to access to all ground, water and air infrastructure. This was specified in the Rambouillet accords, which granted NATO "unrestricted passage and unimpeded access throughout the FRY including associated airspace and territorial waters."[64]

Balkanization driven by the desire to map and code zones to suit military objectives finds a parallel in the 1973 plan outlined by NATO and its Committee on the Challenge of Modern Society, the agenda of which was to globally chart the movements of all commodities and people.[65] Perhaps the most important claim made about this plan is that it was aiming to eliminate any sense of distinction between the civilian and the military.[66] This plan has now been put into operation, with the First Security Forum taking place in Belgrade in 2011 as part of building a broader understanding of the present and future security architecture of the EU. It expects civilian and military relations to become part of a larger reform that would require complete interoperability with the EU and NATO if Serbia is to join the EU. The question of human security has meant that the distinctions between the military and the civilian spheres have been eliminated, since violence and threat are ever-present and never-ending. The citizen body has become conflated with the military body; in this instance, balkanization is not a method of fragmentation, but a merging of what has hitherto remained separate and distinct from each other. It is synonymous with the extension of military control into every facet of society.

This is a different implementation of power from the one exercised during Milosevic's era, which had the territorial agenda of extending the borders of Serbia, rather than erecting architectural structures. Architecture was not accorded particular significance; Milosevic commissioned only two structures during his rule: the (unfinished) underground metro-train station and the Eternal Flame memorial. The current implementation of power is signposted with the vision that the rebuilding of Belgrade becomes the means by which to implement a Western-style legal and economic system aimed at constraining and controlling the movement of people and resources. With the introduction of various formation programmes in the name of security, urban infrastructure becomes the means by which the Euro-Atlantic community seamlessly observes, circulates and extends violence and control. Here, surveillance and disciplining operate beyond moral evaluation; the operation is driven by the attempt to eliminate any tension to do with data and information.

One of the more recent modes of regulation is evident in the control of data during travel. Recent news is that by 2020–21, all citizens from Western Balkans countries will have to submit an online form as part of a European Travel Information and Authorization System before their arrival in any EU country, stating the purpose of the visit and address during their stay.[67] While one may argue that this automated IT system is not discriminatory towards Western Balkans since it will apply to all non-EU passport holders, the very fact that its current planning policy is extended towards this region signifies the assumption that the Western Balkans will not complete the process of accession by 2020–21. In other words, the political agenda of Western Balkans will remain, and the process of Europeanization will remain tied to the transitional politics of contractual regulations. The practical motive behind this legal requirement is not only to map and monitor people's movements ahead of time, but also to presume that Western Balkans citizens are somehow an immediate threat to national security. The perception of the Western Balkans as both an *abjected* zone, and an area of *abjection*, makes it possible to bring in this regulation. The Western Balkans is thus not just about a transition of territory from geopolitical to political entity, but also about the politics of data imperialism, as NATO's Operation has facilitated this reconnection, reorganization and reterritorialization of Serbia. Infrastructure in the name of security is legitimately being deployed in order to observe and make known that which was unknown as well as to retain the violent Balkanist image of the Western Balkans.

Networks of control: energy and water

Admission to the Western alliance comes with a set of network-forming relationships. A 2005 Energy Community Treaty between the EU and nine contracting parties, including Serbia, stipulates the need for all parties to comply with the Treaty and to align with the EU *acquis* and EU 2020 objectives. One of the requirements is a shift of information transparency in regard to each country's emergency oil reserves:

> The level of emergency oil stock reserves is classified as a state secret according to the Serbian Law on commodity reserves. A new Law on commodity reserves is being prepared. Under the new law, mandatory oil stocks would be reclassified as a mandatory stock rather than a commodity reserve, so the level of these stocks would no longer be a State secret. No information has been received so far on the actual level of oil stocks.[68]

It is not disciplining that is of supreme importance here but the need to make all forms of the unknown transparent in order to map, survey and homogenize.[69] While the Energy Community was created to transcend the territorial borders of contracting parties, certain European countries have greater levels of control than others to stipulate conditions on Contracting Parties. That the key aspect of the Treaty is for all the Western Balkans to join EU's electricity grid is telling of an uneven dialogue premised on divesting all capacity of power from newly admitted parties. Integration is a one-way dialogue and a part of political coercion that allows Western countries to retain their dominance in decision-making, despite a surface reading of the Agreement suggesting the abolition of all geopolitical and economic borders. The Treaty is a way to extend control over those areas in the Balkans that have resisted the EU and NATO hitherto.

Natural resources are an important aspect of Serbia's supposed normalization into becoming a 'democratic' member of the Western alliance. Despite the environmental damage done there, this territory continues to make a notable contribution to the rich biological diversity of the Western Balkans.[70] The scientific surveys indicate that this region hosts "[m]ore than a third of all European flowering plants, about half of the fish species, and two-thirds of the bird and mammal fauna."[71] Having access to these resources is significant, so much so that the European Central Bank is investing in the region's energy, transport, water and sanitation, amongst other projects. With Kosovo having 15 million tonnes of brown coal reserves (less than 1 per cent of it mined), and Serbia having over 400 springs of clean and drinkable water (only one-fifth of which is exploited for domestic and international consumption), investment by foreign investors is dependent upon privatization of these resources. The rebalkanization is also seen in the privatization of telecommunications companies, construction firms, banks, power plants, refineries and shipping concerns. The rapid introduction of features such as the privatization, access to information and infrastructure just mentioned exposes the interconnection between the military, corporate capitalism and global politics.

Security and defence increasingly are becoming the means by which to monitor, regulate and control people and territories. However, the manner of their roll-out no longer reflects Foucault's writings on governing as a disciplinary technique of power and control. Now, implementation is effected within the broader network of global surveillance whereby, through the introduction of various reformation programmes in the name of security, urban infrastructure becomes the means of seamlessly observing, circulating and extending control. As a result of current wars waged by deploying surveillance and information operations whose infrastructural networks are globally interconnected, infrastructure has become a direct material and agency through which to execute control and counter control. This was evident during the 1999 targeting of Belgrade, when the prospect of "cutting Yugoslavia's Internet connections was raised at a NATO planning meeting. . . [although] these options [were] rejected as problematic."[72] Considering that the internet is woven into every facet of contemporary societies, it is quite likely that severing the internet connections in Serbia would have had detrimental effects on the rest of Europe and the world. The infrastructure of wires, satellites and unmanned drones thus also comes with inbuilt vulnerabilities.

Control of circulation is not only a matter of incorporating various modes of infrastructure such as water and energy into the global system in the ostensible cause of Serbia's democratization and compliance with EU policies. It is that compliance is what making resources available for purchase is contingent upon. The incorporation is underpinned by the EU's contractual control in an attempt to remove any sense that there are alternatives to the one seen as valid. There have been historical examples of attempts at preempting insurgency in a city, such as the Haussmannization of Paris when, in the name of sanitation and hygiene, planning policies influenced by military tactics were introduced for purposes of surveillance as well as to create a relatively architecturally and economically homogenized city centre; (in)directly the possibility for protests to be staged by the repressed and underprivileged sector of the society was rendered impossible. What is specific to Belgrade at the juncture of hygiene, sanitation and flattening of all sense of Balkanist insurgency is that access to Serbia's infrastructure and resources is one of the key elements in incorporating Belgrade and Serbia as a part of a globally connected and controlled Western network.

Historically, there is a connection between the military, canals and water in Belgrade, beginning during the Roman Empire (44 BC – 1453 CE).[73] During this period, a border line was created along the Danube known as the *limes*, with Singidunum (now Belgrade) acting as both a strategic zone of defence and a transport route for commerce to expand the empire's territorial rule.[74] While the etymology of the term *limes* indicates a limit, it also implies a barrier or threshold that can be bent and stretched, thus making it appropriate to describe Singidunum as an elastic zone and a border between the largest Roman army camps in the Upper Moesia and the supposed barbaric zone known as the Lower Moesia.

With the territorial expansion of the Roman Empire to include Lower Moesia, a shift occurred in the interpretation of Singidunum. The city's role shifted from that of a strategic military stronghold to one of a zone of security and hygiene for the privileged class, prompted by the construction of its first water system. This system spanned several kilometres, from the area of Mali Mokri Lug, which in 85 AD was also a key defensive military camp along the Danubian *limes*, to the Kalemegdan Fortress in the Old City. Belgrade's first water canals measured 1400–1600 millimetres in height and 700–800 millimetres in width, and were built using tiles and stone accessed from the nearby Tasmajdan mines (which at various points in history have been used as burial grounds and shelter).[75] These same water canals were used in 1801 by the Ottoman Janissaries[76] as a transport route to enter the Kalemegdan Fortress and kill the Turkish Pasha Mustafa. The death of the Pasha foreshadowed the First Serbian Uprising against the Ottoman Empire in 1807. As Belgrade became part of the Occident in the 19th Century, the Austro-Hungarian *Bulbulder* water supply was established and added to the existing Roman canals. *Bulbulder* spanned zones from the river Danube to the current day areas of Palilula and Tasmajdan.[77] Later in the century, with Serbia's formation into a nation state, an urban plan by Emilijan Josimovic prompted the construction of the modern water system. Until then, the southern and western areas of what is now Belgrade depended on numerous wells and water springs, known as the *Varoski* system.[78]

While it is undeniable that the history of Belgrade's water systems was tied to a merging of death, violence, security and hygiene, today's investment in water networks is notably more cunning. Water is used as an undulating space for proliferating surveillance, consumerism and control. Since 1999, Belgrade's waterways have become subordinated to the processes of political control. That this city at the juncture of the Sava and the Danube is also at the intersection of two Pan-European Corridors might be the reason behind the significance of waterscapes. The location of infrastructure along the waterways as a precondition for Belgrade and Serbia's admission to the EU exposes the intent to force Belgrade to become a thoroughfare in the world of global connections. The evidence is present in the construction of the Danube waterway Corridor VII and Pan-European Corridor X. Both corridors link ten European countries using the navigable section of the Danube River and are a part of a larger network of ten Corridors. Corridors I–X are part of the Maastricht Treaty of 1992, which aims to facilitate movement of goods and people along routes that range from roads to waterways, railways, telecommunications and energy networks.

The routing of Danube waterway Corridor VII and the almost parallel Pan-European Corridor X is significant, as both would provide connections between West and East. The construction of the 69-kilometre Belgrade bypass connecting to the Pan-European Corridor X (a highway and railroad that runs Salzburg-Ljubljana-Zagreb-Nis-Skopje-Veles-Thessaloniki) has been in planning and a part of Belgrade's general urban plan for several decades. However, the fact that the construction is now largely funded by loans from the EBRD, European Investment Bank and the World Bank confirms that reconstruction assumes not only dependency but also the permanent presence of NATO; any NATO intervention is followed by a proliferation of IMF and World Bank contracts. These various debt-establishing agencies spout the rhetoric of development to disguise the rebalkanization of the Western Balkans by establishing an oppressive system of dependency for this region. In terms of economics, the value of Corridor VII can be seen in the fact that in the period 2001–09, approximately two-thirds of the total foreign company investment in Serbia was directed towards areas along the Danube and its adjacent transport routes.

The political agenda is oriented towards controlling the historically complex Balkans, a zone that over the past 200 years has been associated with a set of behaviours and values that do not conform to those of the West or the East. According to Todorova, prior to the Kosovo war in 1999,

> the dominant paradigm applied to the Balkans translated into the practical ghettoization of the region. The pre-Kosovo European Union visa regime accepted Central Europe but not the rest of Eastern Europe and the Balkans, where restrictions were placed on the movement of populations. This was 'balkanism' in action.[79]

The global agenda under the pretext of transition is used as a means to implement policing and control in less detectable, yet nonetheless disciplinary, ways. Also, it is not simply a question of it being used to spread the market economy as the norm, but that this norm is also used to block any tendency to think of alternatives to the one professed. The paradox of global politics is that while it facilitates access to information, it also obstructs the possibility to propose a counterculture. Infrastructurally, an example of such an alternative during Titoism was the Highway of Brotherhood and Unity (*Auto-put Bratstvo i Jedinstvo*), the transport route linking Ljubljana, Zagreb, Belgrade, Nis and Skopje. This traffic artery that stretched from Slovenia to the Greek border was imagined as infrastructurally unifying and bringing together all its nationalities and ethnic groups.

Work on the road was done predominantly by the Youth Labour Brigades (*Omladinska Radna Akcija* – ORA) volunteers. After WWII, the ORA was established for several purposes: strengthening friendship and solidarity among peoples, spreading Tito's ideology, helping to rebuild the SFRY and even increasing the literacy of Yugoslavs by providing educational facilities within the brigades. Perhaps the most significant point was that the Yugoslav balkanization integrated the youth into the development of cities. Now the same transport route is known by different names, depending upon which former Yugoslav city it is passing through; in New Belgrade, the highway has been renamed the Boulevard of Arsenija Carnojevica. Since 1999, this road has taken on particular importance, as it forms a part of E75, an important international E-road network developed by the United Nations Economic Commission for Europe, connecting Finland on one end and Greece on the other. The Boulevard of Arsenija Carnojevica also connects to the Pan-European Corridor X.

The intent to civilize the region is also seen in the impetus to affix the values and structures of the EU and Anglo-America. This has become particularly evident in the aftermath of the military operation, as remaking a city and a country becomes implicit with a forceful introduction of standardized economic and financial practices. Remaking is contingent on destruction rather than on reconstruction of alternative economic practices. In Belgrade, reformation was seen in the privatization of various telecommunications companies, construction firms, banks, power plants, refineries and shipping concerns. National banks such as Jugobanka, Investbanka and Beobanka were driven into bankruptcy, a move orchestrated by the government but supported by the World Bank and the IMF. Foreign loans were possible only subject to the government facilitating the bankruptcy. The banking void was filled with foreign banks opening their branches in Serbia, bringing with them ideas and practices at odds with what was prior to the 1990s a socialist system. Bringing democracy to Serbia is nothing other than the opportunity to put post-socialist spaces and institutions on the market and proliferate an ever-expanding market economy through privatization, consumerism and security rights.

Reversibility of resistance and violence: transport networks, international loans and DU

With the reformation of Belgrade's transport routes, where priority is given to the Pan-European Corridor X and the Danube waterway Corridor VII, Vuk's Monument[80] underground train station – located in the immediate vicinity of Tasmajdan – offers an 'escape' and a mode of resistance, as it has only ever been partially completed. While the train station was completed in 1995, the metro is still no more than a design on paper. The architect Mirjana Lukić designed the station in 1990. The design and planning occurred during the early phase of Milosevic's presidency and signified the first step towards a mass underground rail and metro system comprising over 30 tunnels beneath the city of Belgrade. However, the realization of the train and metro network in its entirety did not eventuate.

While Vuk's Monument train station was designed to handle 15,000 people per hour as part of a dynamic underground rail network, in 2005, only 10,000 people daily passed through the area.[81] This fragment of the incomplete underground train and metro network includes several levels: an entrance level, passageways catering to multiple uses from a police station to retail shopping, a concourse level, services floor and the station platforms themselves.[82] When the train station was opened on 7 July 1995 (a date that marks the anniversary of the Serbian uprisings against the Ottoman Empire in the 19th Century), it was the deepest underground train station in Europe at 39 metres below ground.[83] The station's depth and its internal bunker-type structure provided city dwellers with shelter during the 1999 NATO strikes.

The underground transport system is at an impasse, and in its unfinished state has an unrealized potential in that, if constructed, it might expose the location of underground tunnels and caves to a wider population. The underground spaces

Figure 5.5 The underground Vuk's Monument train station.

of the city centre of Belgrade constitute a network of interwoven caves, canals, tunnels and circulation routes. Some of these underground spaces, which have remained largely unexplored (their networks are still not known to most people), were used as shelters during various wars and for transport/storage of goods such as food and gunpowder. Almost all buildings in Belgrade's Old City centre sit on underground water bodies, often posing problems for the construction of stable building foundations. It is estimated that 160 water passages flow beneath the greater metropolitan area.[84] Their instability is resonant of Balkanism.

The underground metro has, in a way, remained incidentally a space of resistance due to its very unbuildability. Set against the network of underground water streams and largely unexplored tunnels, Belgrade has a history of unrealized plans for an underground transport system. This planning was subject to many interruptions. The first metro was approved in 1883. The approval covered an area from the dock on the river Sava to Knez Mihailova Street. Other notable metro proposals include the 1958 design for the areas of Kalemegdan, Terazije, Slavija and Cubura.[85] The 1976 iteration was based on five metro lines, while the 1981–82 proposal was meant to establish a network connecting Belgrade from north to south and from west to east. The US$700 million cost of the last proposed system was to be funded from part of the Soviet Union's WWII reparations to Yugoslavia.[86] In the face of objections to this plan from the former Republics of Slovenia and Croatia, construction did not proceed.[87]

In 2010, a plan was put forward to continue the interrupted construction of the proposed Vuk's Monument metro section, this time funded by the French government as part of unpaid WWII reparations to Yugoslavia. However, since these reparations were directed to Belgrade when it was capital of the SFRY before its disintegration, a complex web of financial, legal and military interests now block the way forward. To this day, and despite many proposals and new governments coming on board, the plan to build it keeps getting deferred. The deferral may also be a result of Belgrade having a history of reconstructions, having been destroyed or heavily damaged over 40 times in the Common Era. In the 20th Century alone, it suffered extensive periods of bombing (1914, 1915, 1941, 1944, 1999). Moreover, from its first Neolithic settlement in 7000 BC, Belgrade has been under the domination of numerous groups: the Illyrians and Thracians, the Celts, the Romans, the Ghepids, the Huns, the Sarmatians, the Slavs, the Avars, the Goths, the Franks, the Macedonians, the Bulgarians, the Byzantines, the Serbians, the Turks and the Austro-Hungarians. Its strategic importance began to take shape after the west-east division of the Roman Empire in 395 CE, when what is now Belgrade became a northwestern border zone in the Eastern Empire of Rome. As each new system and power takes hold and begins regulating and coding, there is a discontinuity and the possibility that some spaces avoid being coded. The notion of discontinuity is a mode of resistance. In reference to Deleuze and Guattari's rhizomes,[88] Virilio writes that the globally interconnected system is the "blueprint for military survival. But everything is always reversible, as in Kafka's 'The Burrow': each escape route becomes another threatening entry. Terror too is built into the rhizome."[89]

Escape routes are those Other spaces that have been exposed to different systems. Belgrade's history has been punctuated regularly by wars, destruction and new beginnings. The discontinuities and changes of rulers have meant that some structural layers of the city have not been coded into the new systems. This is not to romanticize any destruction or violence, but to think discontinuity as a possibility to either escape the imposition of control or stiffen it. The history of post-WWII Yugoslavia was an example of overthrowing the controls of fascism, while the aftermath of 1999 is an example of fascism purified and the contraction of control.

While the unfinished underground transport system may be seen as a mode of resistance, there is a complex narrative to do with the significance of the relationship between underground transport networks, economics and violence. It is reflective of Virilio's smooth space which is related to speed, the military and the hegemony of control found in the militarization of space through miniaturization of equipment. In terms of Serbia, this has been implemented through infrastructure and various trade agreements. One way in which the ground was prepared was the economic sanctions imposed on Serbia and Montenegro during the 1990s. However, the sanctions did not have the effect the West desired; instead, they set up the opportunity for Milosevic to strengthen his political rule through the rhetoric that the world is against Serbia and Montenegro and that survival is a heroic trait and historical legacy of the Serbs.

The sanctions opened up opportunities for the black market economy through illegal and unofficial transborder trade. To survive during the economic and travel sanctions meant opening up trade and travel with the East. After 1999, normalization has become contingent on a collective historical amnesia; Belgrade and Serbia are highly receptive to a lifestyle dependent on international bank loans, and they accept risks similar to those of the early 1980s when the SFRY experienced significant degradation of living standards as a result of the inflation driven by post-WWII international bank loans. Such risks have already materialized, with Serbia slowly losing its middle class, and two distinct classes emerging – the privileged and the underprivileged.

The current militarization of space is not as direct as the targeting of Belgrade in 1999 or the violence that occurred prior to and during 1998–99 between the Serbs and Albanians in Kosovo. Violence after 1999 has been more obscured, facilitated not only by international agencies, but also by local moguls and the corrupt Serbian government. Here, Bauman's words find resonance:

> [the] state washes its hands of the vulnerability and uncertainty arising from the logic (or illogicality) of the free market, now redefined as a private affair, a matter for the individuals to deal and cope with by the resources in their private possession.[90]

While fashion has driven a greater level of heterogeneity in New Belgrade's current built landscape, it may still be understood as expressing an economic polarization fiercer than in the preceding socialist context. The military intervention accommodated the introduction of consumerism; needless to say, militarization is ongoing since the culture of debt is largely made possible by IMF and the World Bank, and which in tandem with NATO play a significant role in the 'redevelopment' of cities and orchestration of supposed global security.

Violence may be understood as having transformed values, using consumerism as the conditioner of a democratic society, despite this type of democracy being appropriated to suit narrow interests and depending on long-term memory loss and a consumer culture of spectacle. Even the understanding of time and its use are reconfigured, as time is rationalized and economized for labour in the spirit of 'progress,' Europe and prosperity. In other words, with the onset of consumerism, the dimension of time in everyday life has been directed largely towards values reoriented to consumerism and the market economy. Working hours have been extended, banks are readily lending money for the purchase of apartments and thereby tying people into the market economy system of debt and shopping is being advertised as a mode of leisure and the new norm.

Obligations to meet norms are rampant within EU policies. In 2000, Western Balkans countries were given the opportunity for EU candidature, subject to their allowing the market economy to proliferate and acceding to every legal, democratic and social condition. However, even if all these conditions are met, there is no assurance that any of the Western Balkans countries will be given membership. Another way of looking at the list of conditions for Serbia to join the EU is that membership will depend on conforming to established modes of discipline and resource exploitation for purposes of debt, and control. Even though Yugoslavia's post-WWII history showcased a cyclical pattern of loans, debt and economic depression, post-1999 dependency has been orchestrated differently, from the perspective that in order to create an appealing domestic market for global investors, the fate of Western Balkans was determined with the Central European Free Trade Agreement in 2006; a means to leave behind Balkanist abjectness – "[a]bjection does not provide a reconciliation of meaning. Instead, the abject is that which seems to confound the possibility of meaning" [91] – and enter the normalcy of the EU. The violence underlying this normalcy is the perpetual crisis of life.

The imposition of regulatory normalization conditions is a carrot-and-stick policy. To superficially inflate the economy within Serbia, it needed to open up to more debt. This is a contemporary exercise of feudalism, where the division between nobility (the EU) and serfs (the Western Balkans) is not only territorial, including an uneven power distribution that arises from foreign acquisition of resources, but also intent on extinguishing any account of life lived as well as death suffered in this territory. Taking into account the imminent effects of DU, deployed amongst other reasons for purposes of bringing market economy to this Balkanist zone, Marshall Berman's thinking finds resonance in the way that the fascism of capitalism feeds "on its own self-destruction."[92] In other words, that in its drive to accumulate capital, it would rather destroy even its servants who contribute towards generating profit than stop that which destroys.[93] Moreover, in the attempt to remove the *abject* signifier of Balkanism through processes of balkanization as creation of homogenous zones, an even greater *abjection* is created with the circulatory forces of DU since DU does not recognize boundaries.

Notes

1 Julia Kristeva, *Powers of Horror: An Essay on Abjection* (New York: Columbia University, 1982), 3.
2 Joyce P. Kaufman, "NATO and the Former Yugoslavia: Crisis, Conflict and the Atlantic Alliance," *The Journal of Conflict Studies* 19:2 (1999).
3 Kaufman, "NATO and the Former Yugoslavia."
4 Kaufman, "NATO and the Former Yugoslavia."
5 While the counter argument may be that course content at the University of Ljubljana was taught in Slovenian, this is not a surprise considering that Slovenians were seen as constituent peoples and therefore not considered a minority. Albanians in Kosovo

were seen as non-constituent peoples and therefore a minority. Affording these rights to minorities is significant.

6 Chomsky, *The New Military Humanism*, 24.

7 Dejan Jović, *Yugoslavia: A State that Withered Away* (West Lafayette: Purdue University Press, 2009), 185.

8 Chris Hedges, "Kosovo's Next Masters?" *Foreign Affairs* 78:3 (May/June 1999).

9 Jens Stilhoff Sörensen, *State Collapse and Reconstruction in the Periphery*, 142.

10 "Roundtable: Francis Fukuyama, Eric Hobsbawm, Edward Luttwak, Timothy Garton Ash, Pierre Hassner and Robert Cooper Discuss Global Order in the 21st Century," *Prospect*, 20 August 1999, <www.prospectmagazine.co.uk/magazine/theglobalorderinthe21stcentury>.

11 Julian H. Tolbert, "Crony Attack. Strategic Attack's Silver Bullet?" (Masters Dissertation, School of Advanced Air and Space Studies, Alabama 2006), 25.

12 William J. Clinton, "Remarks at the Legislative Convention of the American Federation of State, County, and Municipal Employees," *The American Presidency Project*, 23 March 1999, <www.presidency.ucsb.edu/ws/?pid=57294>.

13 United States Institute of Peace, "Chapter 4 Economic Issues - Article I-1."

14 Tolbert, "Crony Attack?" v.

15 Benson, *Yugoslavia*, 170–171.

16 Matjaž Klemenčić and Mitja Žagar, *The Former Yugoslavia's Diverse Peoples: A Reference Sourcebook* (Santa Barbara: ABC-CLIO, 2003), 335.

17 Ian Jeffries, *The Former Yugoslavia at the Turn of the Twenty-First Century: A Guide to the Economies in Transition* (London: Routledge, 2002), 604.

18 Tolbert, "Crony Attack?" 34–35.

19 Arkin, "Smart Bombs, Dumb Targeting?" 50.

20 Vojin Joksimovich, "Militarism and Ecology: NATO Ecocide in Serbia," *Mediterranean Quarterly* 11:4 (2000): 140–160.

21 Mlađan Dinkić, ed., *Final Account Economic Consequences of NATO Bombing: Estimate of the Damage and Finances Required for the Reconstruction of Yugoslavia* (Beograd: Stubovi Kulture, 1999), 23.

22 Cited in Der Derian, *Virtuous War*, 197.

23 Stephen Graham, "Postmortem City: Towards an Urban Geopolitics," *City* 8:2 (2004): 179.

24 Demodernization is a term developed by Stephen Graham. For him, demodernization includes the development of tools and technologies that can disconnect, de-modernize and immobilize adversary societies at any time or place deemed necessary. See Graham, *Cities Under Siege*, 272–274, 277–279, 282–283; and Stephen Graham, "Demodernizing by Design: Everyday Infrastructure and Political Violence," in *Violent Geographies: Fear, Terror, and Political Violence*, ed. Derek Gregory and Allan Pred (New York: Routledge, 2006), 321.

25 World Health Organization Report, *Depleted Uranium Sources, Exposure and Health Effects*, April 2001, 45 <http://apps.who.int/iris/bitstream/handle/10665/66930/WHO_SDE_PHE_01.1.pdf?sequence=1&isAllowed=y>; World Health Organization Report, *Report of the World Health Organization Depleted Uranium Mission to Kosovo – Undertaken at the Request of the Special Representative of the Secretary-General and Head of the United Nations Interim Administration Mission in Kosovo* (UNMIK), 22 to 31 January 2001, <http://apps.who.int/iris/handle/10665/108666>.

26 UNMIK, *Report of the World Health Organization Depleted Uranium Mission to Kosovo, Undertaken at the Request of the Special Representative of the Secretary-General and Head of the United Nations Interim Administration Mission in Kosovo* (UNMIK), 22 to 31 January 2001, <http://apps.who.int/iris/bitstream/handle/10665/108666/E71919.pdf?sequence=1&isAllowed=y>.

27 Office of the United Nations High Commissioner for Human Rights, "Part III Methods and Means of Warfare Combatant and Prisoner-of-War Status: Section I Method and Means of Warfare: Article 35(3)," in *Multilateral: Protocol additional to the Geneva Conventions of 12 August 1949, and relating to the protection of victims of international armed conflicts (Protocol I) (with annexes, Final Act of the Diplomatic Conference on the reaffirmation and development of international humanitarian law applicable in armed conflicts dated 10 June 1977 and resolutions adopted at the fourth session)*, 21.
28 Arkin, "Smart Bombs, Dumb Targeting?" 48.
29 Amnesty International, "Federal Republic of Yugoslavia (FRY)/NATO: 'Collateral Damage' or Unlawful Killings? Violations of the Laws of War by NATO During Operation Allied Force," 5 June 2000, 18–19.
30 Peter Sloterdijk, *Terror from the Air* (Cambridge, MA: MIT Press, 2009), 29.
31 Lavrence, " 'The Serbian Bastille,' " 34.
32 The Red Army was formed after the 1917 October Revolution in Russia.
33 Lavrence, " 'The Serbian Bastille,' " 35.
34 Zoran Lj. Nikolić and Dr. Vidoje D. Golubović, *Beograd Ispod Beograda* (Beograd: Sluzbeni List, 2006), 137.
35 Nikolić and Golubović, *Beograd Ispod Beograda*, 111–112.
36 Nikolić and Golubović, *Beograd Ispod Beograda*, 113–114.
37 Nikolić and Golubović, *Beograd Ispod Beograda*, 114.
38 George Monbiot, "Consigning Their Future to Death," *The Guardian*, 22 April 1999.
39 UNEP/UNCHS, *The Kosovo Conflict, Consequences for the Environment*, 1999, 35.
40 UNEP/UNCHS, *The Kosovo Conflict, Consequences for the Environment*, 35.
41 UNEP/UNCHS, *The Kosovo Conflict, Consequences for the Environment*, 47.
42 Zorka B. Vukmirović, "Release of Hazardous, Toxic and Cancerogenic Substances into the Atmosphere During the Warfare in Yugoslavia and an Assessment of Their Environmental Impact," *Environmental Studies, Alternative Academic Educational Network* (Belgrade, Updated 26 October 1999), 4. Also see: Michael Heylin, "NATO Bombs Take Out Chemical Complex," *Chemical and Engineering News* 77:19 (1999).
43 Peter Anderson, "Air Strike: NATO Astride Kosovo," in *The Kosovo Crisis: The Last American War in Europe?* ed. Tony Weymouth & Stanley Henig (London: Reuters, 2001), 200–201.
44 Wim Zwijnenburg and IKV Pax Christi, "Hazard Aware: Lessons Learned from Military Field Manuals on Depleted Uranium and How to Move Forward for Civilian Protection Norms," Report, Netherlands: PAX, 2012, 12, <www.bandepleteduranium.org/en/hazard-aware>.
45 Ruth Yarrow, "The Continuing War in Yugoslavia: Environmental Effects," *Peacework*, September 1999.
46 George Monbiot, "Nato's Dirty War," *Guardian*, 22 April 1999.
47 Joksimovich, "Militarism and Ecology," 140–160.
48 Andreas Laursen, "NATO, the War over Kosovo, and the ICTY Investigation," *American University International Law Review* 17:4 (2002): 778.
49 Paul Virilio and John Armitage, "Ctheory Interview with Paul Viirlio: The Kosovo War Took Place in Orbital Space," *ctheory*, 2000, <https://journals.uvic.ca/index.php/ctheory/article/view/14599>.
50 Sloterdijk, *Terror from the Air*.
51 Chomsky, *The New Military Humanism*, 65.
52 Adrian Lahoud, "Introduction. Post-Traumatic Urbanism," in *Post-Traumatic Urbanism: Architectural Design* 80:5, ed. Adrian Lahoud et al. (London: Wiley, 2010), 19.
53 Paul Polansky, *Unleaded Blood* (Kosovo: Kosovo Roma Refugee Foundation [KRRF], 2005).

54 Paul Polansky, *Unleaded Blood.* Also see: Rick Gladstone, "Roma Sickened in U.N. Camps Are Still Waiting for Redress," *New York Times*, 18 April 2017; Austin Ramzy, "U.N. Offers Regret but No Compensation for Kosovo Poisoning Victims," *New York Times*, 26 May 2017.

55 Grubacic, *Don't Mourn, Balkanize!* 147.

56 Tom Sandler and Keith Hartley, *The Political Economy of NATO: Past, Present, and into the 21st Century* (Cambridge: Cambridge University Press, 1999), 38. Note: the late 1960s coincided with the impending economic changes and introduction of neoliberal economies, while the 1990s coincided not only with the end of the Cold War but also the fall of Communism. The significance of economics and the military is apparent in NATO's 1999 intervention.

57 Economist Intelligence Unit, "Country Report: Yugoslavia (Serbia-Montenegro)," 2000.

58 Cited in Christopher Coker, *Humane Warfare* (London: Routledge, 2001), 152.

59 NATO, *Helping Serbia Dispose of Stocks of Surplus Munitions*, 20 August 2013.

60 Naomi Klein, *The Shock Doctrine: The Rise of Disaster Capitalism* (London: Allen Lane, 2007).

61 Cited in Edward S. Harman, "The Political History of Sham Justice: Carla Del Ponte Addresses Goldman Sachs on Justice and Profits," *coldtype.net* (2005), 3.

62 See Diana Johnstone, *Fools' Crusade: Yugoslavia, NATO and Western Delusions* (London: Pluto Press, 2002).

63 Sten Rynning, *NATO Renewed: The Power and Purpose of Transatlantic Cooperation* (New York: Palgrave Macmillan, 2005), 69.

64 United States Institute of Peace, "Appendix B, Article 8."

65 Paul Virilio, *Popular Defense and Ecological Struggles* (New York: Semiotext(e) Foreign Agents Series, 1990), 64. Note: original text is in italics.

66 Virilio, *Popular Defense and Ecological Struggles*, 90.

67 European Union External Action, "Security Union: A European Travel Information and Authorisation System – Questions & Answers," 5:11 (17 November 2016), <https://eeas.europa.eu/headquarters/headquarters-homepage/15205/security-union-european-travel-information-and-authorisation-system-questions-answers_en>.

68 Energy Community Homepage, *Chapter 15: Energy – European Commission Opinion on Serbia's Application for membership of the European Union 12 October 2011*, 83.

69 Michel Foucault, *Security, Territory, Population: Lectures at the Collège de France 1977–1978*, ed. Michel Senellart (New York: Palgrave Macmillan, 2007), 10.

70 UNEP/UNCHS, *The Kosovo Conflict, Consequences for the Environment*, 1999, 63.

71 UNEP/UNCHS, *The Kosovo Conflict, Consequences for the Environment*, 63.

72 Stephen Graham, "Disruption by Design: Urban Infrastructure and Political Violence," in *Disrupted Cities: When Infrastructure Fails*, ed. Stephen Graham (New York: Routledge, 2010), 124.

73 Vladimir Bunjac, *Jedan Vek Beogradskog Vodovoda i Kanalizacije 1892–1992* (Beograd: BMS, 1992), 11.

74 Mihailo Maletin and R. Novaković, *Beogradski Mostovi* (Beograd: Arhitektonski Fakultet, 1988), 40.

75 Bunjac, *Jedan Vek Beogradskog Vodovoda i Kanalizacije 1892–1992*, 11.

76 During the Ottoman Empire, Serb boys were taken away from their mothers and trained by the Ottoman Army to form part of a millitary squad called the Janissaries. They would return to areas of the Balkans as grown men trained to kill with no recollection of their heritage.

77 Bunjac, *Jedan Vek Beogradskog Vodovoda i Kanalizacije 1892–1992*, 19.

78 Bunjac, *Jedan Vek Beogradskog Vodovoda i Kanalizacije 1892–1992*, 20.

79 Todorova, *Imagining the Balkans*, 191.

80 The monument is to Vuk Stefanovic Karadzic who was a late 18th–early 19th Century Serb linguist and philologist who reformed the Serbian language by using strict phonemic principles.
81 N. C., "Podzemlje za ponos," *Glas Javnosti*, 7 July 2005.
82 Mirjana Lukić, "Javni Prostori u Vise Nivoa Zeleznicke Stanice 'Vukov Spomenik' – Petnaest godina posle," *Izgradnja* 65:1–2 (2011): 70–74.
83 Lukić, "Javni Prostori u Vise Nivoa Zeleznicke Stanice 'Vukov Spomenik,'" 74.
84 Bunjac, *Jedan Vek Beogradskog Vodovoda i Kanalizacije 1892–1992*, 56.
85 Rajko Lj. Korica and Danilo S. Furundžić, "Metro or Light Rail: Belgrade Transport Proposals," in *Proceedings of Real Corp 2011 Conference on Change for Stability: Lifecycles of Cities and Regions*, 18–20.
86 Uroš Komlenović, "Belgrade Metro: Lightly Made Promises," *Vreme News Digest Agency* No 300, 5 July 1997, <www.scc.rutgers.edu/serbian_digest/300/t300-6.htm>.
87 Komlenović, "Belgrade Metro: Lightly Made Promises."
88 Rhizomes do not have an origin or an end and are suggestive of multiple connections in motion, meaning that anything can be connected and reconnected; the rhizome is affirmative as it is about multiplicity and heterogeneity.
89 Paul Virilio and Sylvère Lotringer, *Crepsular Dawn* (New York: Semiotext(e), 2002), 140.
90 Zygmunt Bauman, "Living Together in an Urban World," in *Cities, War and Terrorism: Towards an Urban Geopolitics*, ed. Stephen Graham (Oxford: Blackwell Publishing, 2004), 118.
91 Norma Claire Moruzzi, "National Abjects: Julia Kristeva on the Process of Political Self-Identification," in *Ethics, Politics, and Difference in Julia Kristeva's Writing*, ed. Oliver Kelly (New York: Routledge, 1993), 137.
92 Marshall Berman, *Adventures in Marxism* (New York: Verso, 1999), 139.
93 Gilles Deleuze and Felix Guattari, *A Thousand Plateaus: Capitalism and Schizophrenia* (Minneapolis: Minnesota University Press, 1988), 231.

6 Territorializations

Transitions, thresholds and transgressions

Shaping space softly

Transitions and historical whitewash: Romani

Balkanization appears to be an ongoing political process in that the Europeanization of all Western Balkans countries requires their full political separation (balkanization) from each other and subsequent recognition of each other.[1] As regards Serbia, an additional condition is that it needs to recognize Kosovo. Deeming this process as necessary rests on the belief that ideals such as peace, democracy and stability can be attained only once all the Western Balkans countries join the EU. However, setting these conditions as a qualifying norm assumes that the zone in question has no history of civilization and likely no history of political culture.[2] The overall political process is manifest with the remaking of this territory. After all, the 1990s balkanization of Yugoslavia, and the subsequent rebuilding of cities as tagged by the EU and NATO's agendas, are for purposes not only of Serbia finally attaining 'civilization' but also for an acquisition of Anglo-American and EU's power made possible through privatization, elimination of social welfare and public good, and negation of difference. The grab for power is particularly visible along Belgrade's waterways and in its new housing developments.

The negation of difference is seen in the drive to expunge the 'abnormalities' associated with the Romani. This has been reinforced in spatial terms through dehistoricization associated with negation of history of settlement and occupation; the presence of Romani zones in Belgrade's history has been largely undermined. At their highest population and footprint in 2005, the Romani settlements in Block 18 (located along the left bank of the river Sava) were a bustling and rich patchwork of inhabitation, with each Romani dwelling constructed of different materials; rubbish took on a new meaning and purpose there as a valued and primary element of construction.

The Romani move from southern Serbia to New Belgrade was prompted by the 1980s economic crisis, spurred, in turn, by Tito's death in 1980 and increasing interest repayments on World Bank loans. Since their initial settlement there, the Romani population grew to become 173 dwellings and 820 people in 2005.[3] The increase in settlement size and population was also due to the wars in

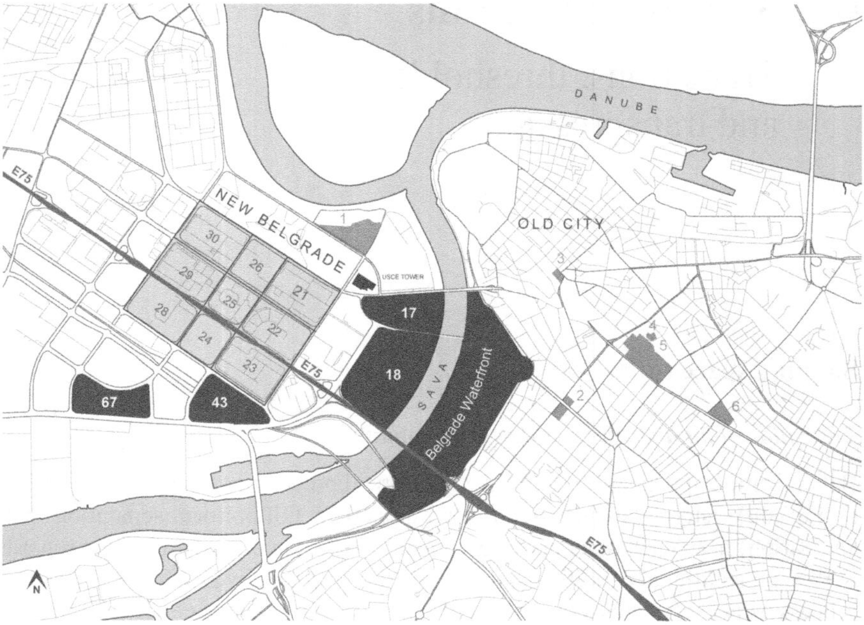

Figure 6.1 The waterfront along Belgrade's foreshore, and New Belgrade: Block 17 – Belgrade Fairground, residences, car markets and small industries; Block 18 – predominantly residences of the Romani and those from lower socioeconomic backgrounds; Block 43 – current day flea market and future bus station; Block 67 – Belville and Delta Shopping Centre; Blocks 21–30 – 1950s urban proposal.

Bosnia-Herzegovina and Croatia in the 1990s, which signalled the disintegration of the SFRY, and the 1998–99 Kosovo War, during and immediately after which 140 new refugee families arrived.[4] The mid-1990s hyperinflation in Serbia further affected the livelihood of some Belgrade locals, meaning that Blocks 17 and 18 grew in size, occupied by lower socioeconomic classes and the disadvantaged. The developments that have sprung up in the past 20–30 years have remained largely unregistered. Settlement size has increased despite the lack of adequate infrastructure: sanitation, sewerage, electricity lines and pedestrian corridors. To address this lack would be the first step towards legalization of these informal and illegal settlements;[5] however, that is not in the interest of the Serbian government considering the prime waterfront position of Block 18.

Besides the fact that Block 18 exemplifies the type of Romani living largely not approved of by the Serbian government (nor, more recently, by international agencies), the central importance of these settlements is that they are located beneath the Gazela Bridge, a major transport route that connects to both the west and the east via E75. This 332-metre-long and 27.5-metre-wide beam-and-arch steel bridge with concrete abutments is, at the moment, the most significant bridge

Figure 6.2 Block 18: the Romani settlements beneath the Gazela Bridge with the transport artery joining E75.

across the Sava. This is due to a number of reasons. Its three lanes in both directions connect Belgrade's city centre with New Belgrade, now coping with more than 40 per cent more traffic than the intended 40,000 vehicles per day when it was finished in 1970. At its opening, Tito was present, perhaps because at the time this was imagined as an artery that forms part of the Brotherhood and Unity Highway, imagined to connect SFRY's Macedonia, Serbia, Croatia and Slovenia.

Although visible to a passer-by only from a distance through the bridge, the settlements established and occupied by the Romani are provocative in that they contradict normative expectations of how 'civilized' and 'respectable' people should live. They do not observe the aesthetic approved by the West or the Serbian government. As they are located in the central zones of New Belgrade, rather than on the fringes, the settlements are difficult to overlook. They expose a normative limit, a signifier of a social and political position of who can live in a city and how that city can be occupied.

In 2007, the European Investment Bank and the EBRD provided funding for the reconstruction of this structurally unstable bridge, but on the condition that the people living underneath it (the Romani) be moved and provided adequate housing.[6] Approximately 50 families (most of them Romani) lived beneath the bridge, all having fled Kosovo in 1999. The requirement to relocate was underpinned by invoking the need for security. The Belgrade city authorities met this obligation by forcibly removing the Romani to various peripheral areas.[7] The options they were given included relocation to housing in tin shanties – a proposal supported by the German government, which donated used tin shanties for purposes of housing through the humanitarian agency Caritas – with no water or electricity laid on, a situation that is admittedly not uncommon in Belgrade.[8]

Needless to say, the sweltering summer months turned these dwellings into furnaces, and in the sub-zero winter months, they became ice blocks. Whilst the presence of the Romani inhabiting the spaces below the Gazela Bridge in Block 18 was perhaps welcomed by Belgrade authorities during the 78-day NATO campaign in 1999, as it saved the bridge from being targeted – just as it appears likely that the rock concert demonstrations staged on Branko's Bridge during Operation Allied Force spared that bridge from destruction[9] – their being there seems not to suit either national or international bodies today. The EBRD's pretext of need for security during the reconstruction of the bridge (which went ahead in 2010–11) was a cunning bid to 'democratically' define how public space can be used and by whom.

Since 2007, the density of the Romani settlements has been reduced significantly, now occupying only the edge of Block 18, as in 1983 when two Romani families settled there despite the immediately adjoining area being used for waste disposal. The 2007 resettlement of the Romani is only one example of violent evictions, serving perhaps as a precursor to the instatement of the City of Belgrade's 2009 "Action Plan for the Resettlement of Shanty [Unhygienic] Settlements."[10] It seems that because of the prevailing absence of a Romani historiography, the violence of these evictions is increasing in tempo. In 2011, another 27 Romani families were forcibly evicted from Block 61, a zone in proximity to the left bank of the river Sava, in order to make space for a new commercial development.[11]

According to Amnesty International, this is momentous, as it is the first time that the state has become involved in questions to do with the evictions of the Romani in Belgrade; prior evictions were carried out by the city authorities.[12]

Considering that the Romani have not had a fixed association with land, territory or a nation state, their very presence goes against the line of thinking that treats land as a commodity with potential for economic surplus and political power, serving as a reminder that alternatives perhaps exist on how to think nation, national rights and legitimization of land and territory. From this perspective, the presence of the Romani constitutes a radical difference. That very little academic writing has been dedicated to the status of the Romani living in this New Belgrade zone since 1983 reinforces the hegemonic position. The relocation of the Romani – the ultimate 'outsiders' – to the fringes from this central location cancels visibility and thinking to do with land, nation and national rights in that the Romani way of life demonstrates that land and territory can be considered beyond the practice of balkanizing enclaves for purposes of exercising control or generating profit through ownership.

Threshold erasures: Belgrade fairground

On an adjoining block, Block 17, a different kind of territorial appropriation is taking place; that of historical erasure of a territory's identification with its first built structure. The 11-hectare Belgrade fairground (*Sajmiste*), located within Block 17, was turned into a concentration camp not long after it was built in 1937.[13] It was used by the Nazis during WWII to exterminate Romani, Jews and all those Serbs who were openly opposed to the Fascist regime. During the period 1947–50, the fairground served as the headquarters for a construction firm that was in charge of building New Belgrade.[14] Some of the pavilions also housed the volunteer members of the ORA who helped lay 8 million cubic metres of sand and gravel as foundation for New Belgrade's post-WWII construction. The fairground was placed under a historical preservation order in 1987. In 1995, a memorial sculpted by Miodrag Popovic was erected in its vicinity to mark the violent extermination of peoples during WWII; while commemorating the manifestation of genocide in Belgrade on the fairground site, the memorial plaque also mentions the Serb and Romani victims of the concentration camp in Jasenovac (Croatia) and marks the heroism and resistance of all Yugoslav peoples and victims of the Nazi terror.

The fairground used so violently by the Nazis during WWII has been layered with more recent violence. Recent proposals include the conversion of the fairground complex into a Holocaust memorial and museum.[15] This adaptation of the complex would entail the displacement of the zone's very heterogeneous community (artists' studios, the residences of the Romani and those from lower socio-economic backgrounds, car markets and small industries). The first evictees were forced out in 2013. The association of this fairground with one particular event reduces the salience of other events as well as other examples of violence that have occurred on this site and within these pavilions.

Figure 6.3 Block 17: new and old residential dwellings and the Belgrade fairground.

This is not to disregard the immensity of scale and extreme nature of violence during the Holocaust; however, associating the fairground with only one violent event fixes history to a frozen image, one that places measure and ordering on violence. It is a portrayal of violence as a fixed and obvious matter that can be quantified and compared. This categorization, whereby so-called quantity or historical singularity becomes the indicator of the extremity of violence, further facilitates negation of other histories of violence. This is the politics of the right to historical voice, an ordering against which all other kinds of violence are measured. Classification of violence prefigures an ordering on who may speak and who needs to remain silent. In more extensive and even spatial terms, it eliminates the opportunity to question the strategy (*how*) and the intent (*why*) behind violence, beyond the voice that has been given the historical and hegemonic right to speak. Here, the meaning of violence becomes predetermined, further restricting the possibility to understand history in any way other than through the lens in which it has been framed.

Blocks 17 and 18, located on the left bank of the river Sava, are geared towards development; this is an extreme transition, from the post-1945 socialist conception of *space for all* to a right-wing, consumer economy and private property–oriented *market for the few*. For Block 18, a new street and urban network has been proposed, treating Block 18 as a tabula rasa. Elimination of this heterogeneity would negate not only the contemporary nature of living and building there but also the historical status of this zone as a unique example of rural dwellings built in the 1930s in an urban context.

The zone along the left bank of the river Sava (present day Block 18) was the first area to be developed, even though no urban plan was approved. It was probably chosen for construction of residential houses due to its slight elevation above the level of the wetland. Prior to the purchase of 200 hectares by the Yugoslav King Aleksandar in the mid-1930s, the land was owned by farmers and used to graze sheep. The zone was divided into seven linear streets (only six streets exist now) and subdivided into 200 long, narrow parcels, sold to individuals and used for the construction of houses. Initial construction on each parcel was primarily of a small, single-storey house on the edge of a strip and closer to the street, with a long yard behind utilized for orchards and vegetable gardens.

The site may be seen as a living archive that shows the transformation of a single-storey residential type over time, with changes influenced by fluctuating social and economic factors. Seemingly, the intent is to whitewash history for purposes of financial gain in that the less history a zone is presented as having, the greater scope there is for profit. Both Blocks 17 and 18 have a uniquely historical position in New Belgrade, the balkanization of which violently negates both historical and current conditions. The violence is largely predicated upon the ironing of history in that it ignores the fact that the erection of New Belgrade began in the area that is today known as Block 17 and Block 18 in the mid-1930s; the development of these areas occurred despite no urban plan being approved, even though a number were proposed since the 1920s.[16]

Privatization and Belgrade's greenbelt

With Serbia's shift to a 'democratic' government in 2000, land privatization also started to occur. In 2003, a new law was adopted under which the City of Belgrade was established as the owner and user of land in New Belgrade, while the Old City remained predominantly in individual private ownership.[17] However, in 2014, with the development of Belgrade's waterfront along the right bank of the Sava, not only were 180 hectares of land there gifted and the ownership transferred to a foreign investor, but the investor was exempted from local taxes and fees. The land along the right bank of the Sava has been converted to private leasehold, to the detriment of public amenities that have been extinguished. Development is largely intent on eliminating any distinctness specific to Belgrade in that there is a flattening of the possibility to activate space for public interests; use of space is instead geared towards private interest in the construction of mixed-use developments and commercial architecture. The waterfront urban redevelopment has become a possibility only as most of the buildings in this zone have been snapped up inexpensively by the Serbian upper middle class from their lower socio-economic owners.[18] They have been acquired largely for short-term ownership; the medium-term agenda is driven by the vision to sell these buildings at a substantial profit, since they fall within the Belgrade Waterfront plan.[19]

With the imminent redevelopment of the zone along the river Sava, the historic green belt along the rivers is also in question. The green belt along the Sava and the Danube was drawn into Belgrade's urban plan with the arrival of the architect and mathematician Emilijan Josimovic in Belgrade. Josimovic's 1867 urban plan proposal brought about the reformation of the city. At the time, the Old City of Belgrade was separated into three zones: the Turkish population occupied the Kalemegdan Fortress and the north areas of Belgrade's Old City sloping towards the Danube, the Jewish quarter was located south of this and the Serb district took up the area sloping towards the river Sava. Josimovic's urban plan for Belgrade was guided by the objectives of regularizing the congested and labyrinthine city streets and creating a more uniform plan to the city.[20] It was perhaps predictable that the 19th Century reconstruction of Belgrade would involve erasure; Josimovic was a contemporary of the famous urbanist Georges-Eugène Haussmann. Unlike Haussmann's 19th Century violent intervention in Paris, however, which attempted to clean out that city and suppress the tendency of the lower economic class to revolt, Josimovic's objective was predominantly to eliminate memory of the centuries-long rule of the Ottoman Empire.[21]

Figure 6.4 View towards the Kalemegdan Fortress and the port along the top right end of the river Sava.

The reshaping of Belgrade as part of a modern nation state was a form of balkanization in the sense that it was an attempt to erase the memory of Oriental traces. Perhaps because Josimovic's plan remained relatively vague about the future status of mosques, the erasure does not seem to have been overly premeditated.[22] Nevertheless, with the general expansion of the Habsburg Empire and the Occident in the mid-19th Century, and the associated change of values, the cultural and religious markers of the Orient were left to decay. Now, a century and a half later, the focus has moved to erasing Josimovic's instatement of the green urban plan along the water bodies, intended for purposes of recreation and, more important, access to green space for all socioeconomic strata. There is an impetus towards eliminating all reminders of Yugoslav socialism in terms of public ownership. It is driven by the agenda of controlled socioeconomic settlement, a far cry from the 19th Century when the Serbian Prince Milos allowed the land along the Sava to be freely settled. In response, the zone was quickly settled by predominantly poor people and the Romani from not only Belgrade but also Serbia more broadly.

While the rest of Belgrade's central zones were developed, this zone has evaded implementation of any rigorous planning visions, and in spite of or because of this, the area is a rich and diverse source of ideas, scales and approaches to building and aesthetics. The argument is that its run-down state is precisely the reason why the area along the waterfront is in such dire need of development, and further that high end development there is appropriate since Belgrade is largely devoid of luxury apartments within the city centre. The rejoinder to that argument is that the increase in population is predominantly along Belgrade's peripheries and comprises people who would not be able to afford such apartments considering that the average monthly wage in Belgrade is approximately £300–350. Also, the value of once public and state-owned greenscapes used for cycling and rollerblading amongst other leisure activities is now being traded to privilege only a particular public (the upper middle class) for whom greenscapes foreground consumption associated with shopping centres.

It is not that consumerism is negative, but that it, in this example, preempts diversity of voices by privileging a particular way of thinking and living. In the attempt to flatten diversity of aesthetics and functions (from physical buildings to use), what also comes into play is the desire to eliminate the alterity and associated tensions that make a city a dynamic place. The repercussions include not only ghettoization of economically disparate areas, but also elimination of social solidarity. In this process, what become obvious is the way in which individual choices are eclipsed if they attempt to deviate from consumerist ones. According to Bauman,

> the art of . . . democratic politics, is about dismantling the limits to citizens' freedom; but it is also about self-limitation: about making citizens free in order to enable them to set, individually and collectively, their own, individual and collective, limits.[23]

He further suggests that the latter category of citizenship is almost extinct as a result of limiting all choices other than those dictated by the needs of the market economy.[24] This is particularly significant for Belgrade considering its history of social awareness and the cohesion that came with the socialist brotherhood and unity tag.

Desocialization and Belgrade Waterfront

While the initial redevelopment of the waterfront Savamala (Little Sava) district in the early to mid-2000s, encouraged by city authorities, was oriented towards this zone becoming a design quarter through development of art and culture,[25] more recent redevelopment as seen in the Belgrade Waterfront 2012 master plan is tied to a gutting of all sense of socioeconomic diversity and alternative design and cultural practices. In other words, more recent developments are nothing short of savage gentrification of a zone now to be associated with a transitional post-socialist identity intent on membership in the EU. What makes the initiation of the Master Plan even more chilling is that, despite the initial talk that all investment proposals would be assessed through a tender process, the only name chosen and announced as investor-builder was the Government of the United Arab Emirates and private international company Eagle Hills. At the time, Prime Minister Aleksandar Vucic, now President of Serbia and president of the Serbian Progressive Party, said that the development would change the image and identity of Serbia; the projection also being that the Belgrade Waterfront would help get Serbia out of (impending) economic crisis. This was the first time that the Serbian government became directly involved in any development; it both initiated and facilitated the project, suggesting that the development was not only of national importance but also that the project was above all laws since it is congruent with the change of the legal constitution. As such, the urban development is playing a key role in the rebalkanization of Belgrade for purposes of consumerism. The process of shedding one identity and adopting a new one is necessitated by economic reforms and overall privatization of industries and resources. To secure this new identity, Belgrade and Serbia must be transformed into a balkanized enclave purely dependent upon a corrupt state government and foreign investors.

This is not just a question of the profit margin, but also that the new global identity is contingent on depoliticizing the historically complex and liminal identity of Balkanism, and in this process, blocking out the possibility for social rights and citizen participation. The 'regeneration,' now associated with the construction of predominantly upper middle class commercial and residential complexes, has been cloaked in opacity, particularly between the government and the citizen.[26] Attempts at silencing the public go so far as simply declaring "[t]he public interest [to be] established for the expropriation of real estate for the purpose of [allocating] the land . . . for the construction of commercial and residential complex 'Belgrade Waterfront'."[27] What has made lack of transparency possible is the elastic manoeuvring of the law for purposes of determining public interest; any proposed legislation is halted if it appears to put the development into question.

Figure 6.5 Savamala (Little Sava) from Branko's Bridge in 2010.

Until one aspect of this project, the proposed shopping centre – the largest one in the Balkans and a signifier of a new face of Belgrade and Serbia – is constructed, some of the green spaces are used to house funkily decorated mobile food carts, a smorgasbord of Middle Eastern, Indian and European cuisine; staging the belief that not only is Belgrade joining the world but also that the world is coming to Belgrade. The Belgrade Waterfront is a reinstatement of brotherhood and unity, in a superficial way. The emerging typology of mobility is a far cry from Foucault's counter heterotopias as spaces that open up and destabilize the governing powers within any city. These mobile carts are emblematic of a new type of power where any sense of openness and heterogeneity is conditional on diversity and culture being subsumed into a spectacle of reductive imagery to represent that culture. To engage with a culture through a single cursory layer corresponds to the complexity of flavours altered to suit the habits of taste buds found within this region, food ready to be eaten on the go.

The city is slowly being emptied of its heterogeneity – of identities, relations, encounters and beliefs – by being turned into a series of enclaves that cater to the largely wealthy minority and (in)directly drive the parameters of how the city may be moulded to suit bigger commercial agendas.[28] The redevelopment of the Waterfront is nothing more than a tested recipe of market investment urbanism and privatization-driven planning. The proposed commercial developments showcase lack of creativity by projecting a new vision for Belgrade through the well-worn commercial image of waterfronts blended with a dense assemblage of skyscrapers and glitzy façades, and all finished off with multiple shopping centres, manicured green spaces and high surveillance. The grand attempt is to remove all association with Titoism; any remnant of nostalgia for that period delays the transition to market economics.

The reductionism of styles and opportunities afforded is also seen in the proposed architecture and public spaces. Recently, the barges on the left bank of the Sava – located directly opposite the waterfront development and within New Belgrade's Blocks 17 and 18 – used for purposes of entertainment (predominantly as restaurants and nightclubs) have been given orders by the local government to either close or relocate. Apparently, the excessive noise was causing residents to complain. That the residents of both Blocks – the Romani, refugees from the 1990s wars in Croatia and Bosnia-Herzegovina, and those from low socioeconomic backgrounds – actually have put forward these complaints is hard to believe. Even harder to believe is that their appeals were listened to. After all, these are *the residents* who have been forcefully relocated to the fringes of the city in 2007 and 2011 or have been forewarned that they will need to relocate.

If architecture and urban spaces are the index value of the society,[29] Belgrade is suggestive of a distance-shrinking and time-compressing consumer culture of noise associated with billboard signs and their symbols, and where more significantly once-exterior spaces are becoming interiorized. The city is turning into a variation of Las Vegas; however, unlike the possibility to engage with and comprehend the buildings offered by travel in a car driving at high speed,[30] the encounter of speed in Belgrade is very different. When travelling along a highway, there is a push to eliminate any sights that transgress against the norm of acceptable living standards, such as the Romani settlements along the left bank of the Sava. When infrastructure is deemed no longer necessary and of use to the government, buildings are pulled down at unprecedented rates, people move out without hesitation or delay and dialogue between citizen and institution are interdicted. In May 2016, a number of buildings located in Savamala – a zone on the right bank of the Sava now utilized by the Belgrade Waterfront – were bulldozed. The residents who were there were tied up, and any attempt to contact the police over the matter was blocked. Overall, the changes in Savamala were rapid in an attempt to remove the possibility for citizens' protest and largely oriented towards closing the city off from any type of engagement that may question the need for privatization of once-public spaces.

Unlike the duck buildings found along the Las Vegas highway strip, whose function can be comprehended only through its symbolic form, ducks found in Belgrade have a different and more hopeful application. The duck is the key symbol of the Don't Drown Belgrade/We are not Handing Over Belgrade (*Ne da(vi)mo Beograd*) initiative. This activist and citizen-initiated group brings together people of different profiles and professions with the aim of raising awareness around the degradation of Belgrade, as seen in the construction of hyper-colossal projects such as the Belgrade Waterfront. The first, bigger public display

Figure 6.6 Part of the Waterfront master plan development advertising campaign at Belgrade's Nikola Tesla airport in 2018.

of dissent was the February–March 2017 street protests. Some of the stunts deployed included Operation Lifebelt (*Operacija slauf*), where through beach toys, songs and noise the message was conveyed loud and clear that citizens need to be included in any decision-making regarding development – for themselves and in the name of Belgrade as well as to allow the greenscapes along

Figure 6.7 Belgrade Waterfront development in Savamala in 2017 (top) and 2018 (bottom).

the riverfronts to remain public and used for leisure activities. Other messages included that public resources need to be injected for the benefit of citizens and social good, rather than purely for investors. While the large protests in public spaces have been scaled down, the online presence of Don't Drown Belgrade/We are not Handing Over Belgrade with its critique of initiatives driving the redevelopment of Belgrade is still present.

The 2018 campaign, in response to the local elections, included stunts from a renovated vehicle in the shape of a duck called duck-mobile (*patkomobil*) that was driven around the city to stage acts such as pulling up to the main entry of the Parliament House in Belgrade, to generating video material in the form of jingles and songs to draw attention to the problems of privatization spurred by investor urbanism.[31] All the stunts have been documented as part of a marketing campaign and shared online to foster broader reach and solidarity, while also incurring lower marginal costs than are normally involved in professionally orchestrated election campaigns. Undeniably, it is very possible that the state mass media networks would have attempted to suppress the campaign considering that the mass media in Serbia is already (largely) censored. Despite the local elections swinging in favour of the Serbian Progressive Party,[32] no surprise considering the number of 'ghost' voters (deceased and non-Belgrade residents), the Don't Drown Belgrade/We are not Handing Over Belgrade campaign has received international support from Ska Keller, co-president of the German Greens/European Free Alliance, and Yanis Varoufakis, the Greek economist, academic and politician.

Balkanism and balkanization: distinctness, diversity and alterity

Historical shifts and mixed-use development

Despite the proposed Waterfront development not being designed with the public interest in mind, and despite the protests that have been mounted by Don't Drown Belgrade/We are not Handing Over Belgrade, the project is going ahead. The development of the Sava-facing waterfront[33] is also driven by the intent to relocate the bus station to New Belgrade's Antifascism Battle Street (Block 43), which currently houses a flea market (*buvljak*).[34] The irony of this relocation is that despite the street name suggesting resistance to extremist right-wing politics, the current remaking of Belgrade is precisely geared towards the attempt to eliminate the battle found in alternative thinking. Even though the reconstruction has been presented as diversifying the palette of opportunities as well as opening up the city and the country to the world, the vision for this *new* Belgrade – one of the last European capital cities to join the market economy in the aftermath of 1999 – is not only driven through policy, but also made operational obliquely through desocialization. This has been seen in the abandonment of the socialist right to a residence and even the corollary understanding of a residence as a possession. Those are now being replaced with a new understanding: residence as a commodity. What is being removed is the history of alternatives that were implemented in constructing New Belgrade, which was designed strategically for purposes of serving as the capital of (what with the 1963 Constitution became known as) the Socialist Federal Republic of Yugoslavia (SFRY).

The urban plan that facilitated these construction priorities was devised in the 1950s. This is significant considering that many urban plans were proposed since the 1920s, though none were implemented. The 1950s plan – constituted of tall block-type architectural forms separated by wide roads – dealt primarily with the central zone of New Belgrade, constituting nine symmetrical blocks (Blocks 21–29) whose central axis was formed by three blocks (Blocks 24, 25, 26) intended for public use and serving as the grand assembly area. The symmetrical blocks fronted the building of the Federal Executive Council (1947–61) and were backed by a proposed railway station, which was not built.

Despite the projected 1950s image of New Belgrade as a centre for government and cultural buildings, it instead developed a residential character as the SFRY experienced great housing demand following the end of the war. Construction of residential accommodation was accelerated as a result of SFRY's socially owned property system, in which state institutions provided housing for their employees. The specificity of the housing function followed the ideological premise that a place of residence in socialism is not a commodity but rather something defined by its use value. This reflected another legal and social standpoint, that of the right to a residence as a fundamental right to the common public good and the well-being of society, and related to the ideal of just distribution, that is, the ideal of a free apartment, and free social services for all.

With residential buildings taking precedence, construction of the central axis was halted. New Belgrade was filled with Brutalist[35] buildings, which were seen as a highly efficient solution to the great housing demand in the aftermath of WWII.[36] The orthogonal Brutalist blocks were planned with hierarchical vehicular/pedestrian circulation. The hierarchy was achieved in three ways: first, by having major streets connecting the area; second, by having an internal street system between the blocks consisting of primary streets; and third, by having bridges connecting the blocks to a pedestrian zone within each block. The plan closely followed the principles of Le Corbusier's Radiant City, Brasilia, and reflected the programme of CIAM, by which strict functionality of urban planning became the social means of improving the living standards of the urban population. Modernization was a signifier of urbanization, industrialization and socialization. The urban construction was part of a larger political project whose motives were ideologically unique and experimental: the interlacing of socialist social welfare with a planned market economy.

With Tito's death in 1980, and the ensuing economic crisis resulting from increasing state indebtedness to the World Bank, the planning, use and decision-making process concerning land developments of many unbuilt zones within designated Blocks also started changing despite the incompletion of New Belgrade's Modernist urban plan.[37] In the 1990s, with the violent disintegration of the SFRY, which also led to the imposition of trade and travel embargoes on Serbia and Montenegro, the state started to sell apartments in New Belgrade to their occupants.[38] It was an attempt to deal with growing hyperinflation flowing from conflict and the newly imposed foreign policy measures. With this shift, an apartment was no longer a social right, but a possession. Since a residence was no longer a right but a possession, time was significantly shifted and oriented towards earning in order to rent or own a property. The notion of common good was now headed towards individual survival.

The privatization process was accelerated after 1999. The rapid reconfiguration of New Belgrade's central axis and other blocks has been driven mainly by international investment, foreign loans, contractual stipulations of the EU, corrupt local Serbian politicians and their just-as-corrupt international clients. This is evident in another zone of Block 67, where, in preparation for the 2007 Universiade Belgrade (25th World University Summer Games), one area of New Belgrade's

Figure 6.8 Block 67 in 2010: Belville and the Delta City Shopping Centre; and the Romani settlements which are now expunged.

'vacant' blocks was turned into an Olympic Village called Belville, with the intention that the 2,100 apartments in the Village would be sold after the event.[39] Whilst this practice has become common in contemporary Olympic real-estate development (evident from Beijing to London), what is specific to the context of Belville in Belgrade is that the construction was preceded by evictions that targeted a specific ethnic group – the Romani. In a period of six years to 2012, about 250 Romani families were relocated to the peripheries and housed in tin containers.

The planning and construction of Belville exposes cultural desocialization, evident in the creation of a new consumer culture. One of the first examples was seen with the construction of the Delta City Shopping Centre (Delta Holdings) located in Block 67; this was the first Western-style shopping centre constructed in Serbia in 2007. Hypo Adria Bank and Delta Holdings funded Belville's construction. It is worth noting that the owner of Delta Holdings was a supporter of Slobodan Milosevic during his rule and therefore likely to have been a NATO target at that time. Presently, though, the old associations appear to have been forgotten as the current leanings are directed towards accelerating the market through consumerism and privatization. Apart from the increased number of commercial properties, there has also been a change in appearance, use and renaming of particular buildings. One example is a building that was used as the headquarters of the Central Committee of the League of Communists of Yugoslavia (also known as the CK Tower) and was one of three significant governmental/cultural buildings in the post-WWII formation of New Belgrade.[40] The 1962–64 100-meter-tall CK Tower, the tallest structure in SFRY when it was constructed, was targeted during the 1999 strikes. It was also one of the first renamed and reconstructed buildings, a process that lasted three years (2002–05). Now occupied by private and commercial offices and known as the Usce Tower, the building is a signifier of the impulse to eliminate socialist practices and introduce a consumer and private culture. The tower has had two more storeys added to its initial 23, primarily used for hospitality and fitness activities, whilst the greenscaped public area surrounding the building was taken over for the construction of what is now known as the Usce Shopping Centre and a car park that opened in 2009.

The green spaces that were once public commons have been turned into private and inert pristine spaces to be viewed and consumed rather than used. Indications that the consumer culture of spectacle now plays a more prominent role are amplified with the reclad glass façade of the Usce complex (tower and shopping centre) turned into a billboard of messages. At night, the volume of the complex hauntingly disappears, apart from the array of spectacle manifested by its neon light signs. The ultimate irony of the reconstruction is that a tower once associated with Communism/Socialism now boasts the sign of the first market economy bank – Hypo Adria – introduced to Belgrade and Serbia after 1999. The global and political restructuring, in which buildings as well as goods are becoming commodities for consumption and which re-educate residents to believe in the necessity of consuming, underscores the proliferation of spatial violence.

This is not to say that every development is an instance of gentrification; only those where the development is largely oriented towards increasing real estate and profit margins, expunging the diversity of uses and socioeconomic profiles. The

Figure 6.9 Usce Shopping Centre in the foreground, and the reconstructed Usce Tower in the background.

increasing real estate and profit margins come about precisely because the world of consumerism feeds on a continual loop of creating, managing and extending the existence and availability of various products, whilst also neutralizing and flattening alternatives since diversity is possible only if it is tied to the profit margin of the market economy. The balkanization of Serbia post-1999, and Yugoslavia during the 1990s, has been oriented towards the accommodation of consumer culture and debt. Speed is a significant factor in this process; the more swiftly the changes are made, the less scope there is for protest.

Gentrification versus balkanization: militarization

The question that arises then is what the difference between gentrification and balkanization may be, since gentrification is contingent on the city authorities redeveloping run-down areas to repopulate newly created zones with the middle and upper classes. Understanding the differences between gentrification and balkanization is particularly pertinent since Neil Smith articulates that gentrification, that is, 'urban regeneration,' is not just pursued as policy by many states in Europe, but also is a preference of the EU.[41] The policy is promoted for the revenue it yields, but at the expense of socioeconomic segregation. Balkanization operates on a different level. It is a two-fold process by which gentrification is achieved, the first being that through the rhetoric of abnormality and lack of order, areas are prepared for a necessary and largely unquestioned 'regeneration.' The second aspect of this process is that balkanization is a part of a broader strategy to homogenize, whether this be socioeconomically or in terms of creating ethnically homogeneous enclaves, as seen in Bosnia's Dayton Agreement. Balkanization, in the former Yugoslav context, is also tied to questions of democratization, peace and stability – or at least the perception that these values are realized.

Implementing peace and democracy is contingent on the violent instilling of discipline and control. Using Michael Hardt and Antonio Negri's thinking, violence is a question of procedure and an ordering enforced for purposes of not only managing social hierarchies but also regulating social life.[42] Thus, to remake a country and rebuild a nation through, first, 1990s economic and travel sanctions and destruction and then imposition of a transitional identity is less reminiscent of the birth of a nation and more akin to colonialism in that colonial powers first balkanize the world and then politically and socioeconomically absorb the newly created zones through human exploitation and resource extraction. The tag of transition, which has marked this region for close to 30 years – since the start of the 1990s – is driven by the agenda of eliminating the alternative practice of Titoism, reinforcing the derogatory tag of Balkanism, as well as an opportunity to open the transitional systems of the former Yugoslavia in general, and Serbia in particular, to urban militarization through political, social and economic policies.

To militarize the urban is to eliminate any form of Balkanist alternatives and prepare a space where any opposition beyond that advocated by Anglo-America and Western Europe is smoothed out. This is a different smoothing from the one found in the writings of Deleuze and Guattari on smooth and striated space, concepts used to show the inter-relatedness of control and counter control. For Deleuze and Guattari, striated space is inclusive, gridded and finite, and smooth space is extensive, open-ended and infinite. It is the tension between zero and one, the relation inclusive of a push-and-pull network, thereby suggesting that within every measured and territorialized space, there are opportunities to resist and reterritorialize, since even the most disciplinary striated space can foster the deterritorializing opportunities of smooth spaces.[43] Smooth space found in the aftermath of 1999 in Belgrade and Serbia by contrast is more reminiscent of Virilio's understanding on smooth space, which is related to speed, the military and the hegemony of control – all of which are found in the militarization of space through miniaturization of equipment.

The militarization of Belgrade is exercised through desocialization, dehistoricization and negation of diversity through policy. This militarization is not comparable to the militarization of cities in the 19th Century or during WWII. Instead, the militarization process is a means of organizing a society in such a way that violence is produced, though concealed, through the reformative agenda of the post-socialist tag. The transition from the undemocratic Balkanist city and zone to become a valued democratic member of Europe and the world is marked with socioeconomic polarization, privatization of land and infrastructure, and elimination of solidarity. More so, the preference for the marginal zone of Belgrade and Serbia to mould itself into the centralized identity of Europe is the opportunity to eliminate any European or NATO act of violence from the start, since acceding to centralization automatically declares acceptance of, first, the derogatory and violent agenda associated with the Balkans and, second, the need to remove the liminal alterity present within this region in the name of attaining Western civility.

Post-socialism through balkanization is ultimately driven towards the establishment of post-Balkanism, a dispensation under which the people who live in the region finally learn to behave and cooperate in ways thought fitting by the rest of Europe and the West. To eliminate socialism and Balkanism is a sign not only of violence permanently made present and spearheaded by military interventions, but also of the erasure of a belief that alternative systems are possible. Such alternatives were certainly present in the former Yugoslav context. They were to be found in the form of a state constitution as a conclave of multiethnic republics including the instatement of workers' self-management,[44] and in the alternative approach taken under Titoism to questions of economics, and of the military. SFRY was one of two European countries – the other being Cyprus – that openly opposed Europe's Eastern and Western Bloc divisions.

For Jean-François Lyotard, violence in the name of Enlightenment values is always enforced by the merchant and/or economist and the bureaucrat.[45] Violence is implemented through *softer* methods; utilizing Foucault's thinking,[46] through the policies and regulations that need to be obeyed in order to establish the truth that this Balkanist region is ever-so-barbaric and therefore in need of discovering obedience. These *softer* methods are nothing other than a sign of colonialism and colonial administration enacted for purposes of accommodating and maintaining supposed peace. Abiding by these new regulations is not a sign of becoming European, but rather that European political and legal standards are being adopted.[47] If anything, the colonial system implemented is oriented towards the Western Balkans – a zone that belongs neither to the first nor the third world – becoming a third-world zone that is administered through European politics.

Balkanization and distinctness: pre and post SFRY

For Srdjan Jovanovic Weiss, the balkanization of 1990s Yugoslavia has provided an opportunity for identity distinction through the creation of more national capitals (from just one main capital Belgrade, to now Belgrade, Zagreb, Sarajevo, Ljubljana, Podgorica, Skopje and Pristina). This fragmentation is seen to accommodate particularity and "territorial specificity."[48] According to Kai Vöckler, it is rather that the hybridization of different aesthetic styles and typological mixes exemplified in instant urbanism, and spurred by transitional politics, demonstrates how to develop the 'European city'; it is a developmental model of Balkanology.[49]

Utilizing Bakic-Hayden's thinking, "territorial specificity" was already present in that SFRY states identified themselves in terms of what "one is not in relation to the other,"[50] the distinction predominantly centring on one republic being less or more developed than another. It is not simply that each capital identity was unique in its building style; they cultivated distinct arts, economic and cultural palettes. If anything, post-1990s distinctness is more suggestive of each capital city being rebuilt with various foreign loans and thus being amalgamated into the homogeneous spectacle of global consumer culture. Weiss' distinctness, and Vöckler's belief in the radical potential in instant urbanism of corporate and private properties, are possible only if thinking distinctness and potential means complying with the global hegemony of the market economy agenda, which for Andrej Grubačić is "balkanisation from above."[51]

Where Weiss sees further positive aspects of balkanization is during the early 1990s Yugoslavia, when the flood of refugees created an alternative way of living and started diversifying the scale and typology of buildings. On one hand, he compared newly created extensions or temporary buildings that encroached over sidewalks or were perched over rooftops as both inventive indices of a will to live and hilarious architecture.[52] Another example of architectural vitality was seen in the construction of "mushroom houses," which were essentially residential upgrades in the form of two vertically stacked buildings, the top being the larger and newer one.[53] This is in sharp contrast to "turbo architecture," which for him was exemplified by either enormous kitsch villas built by shady businessman or commercial buildings.[54] What distinguished these massive buildings is the unorthodox and clumsy melding of quasi-Byzantine architecture with neoclassical elements to form lavish castle-like structures. Turbo architecture, much like turbo folk music,[55] enmeshed many different stylistic influences during the heights of corruption, criminality and Milosevic's nationalist rule. All of them were built without planning permissions and marked the height of urban and architectural chaos. It is said that the construction of about 150,000 building shells occurred during the period of Milosevic (1989–2000).[56]

Figure 6.10 CIAM-inspired housing developments in the background, and the 'temporary' structures constructed in the 1990s. These structures are still in use.

Illegal construction is a result of not only systemic failure among urban regulation bodies but also the general economic and political isolation. Notably, the organized crime and illegality present during Milosevic's reign has not stopped, even after the supposed democratization of Serbia from October 2000. The violence of *soft* methods, evident in the bankruptcy of national banks and subsequent foreign ownership privatization, and the expunging of the presence of Romani and people of lower socioeconomic background under the pretext of security and regeneration, all manifest the presence of criminality. Now, however, it is portrayed as legitimate, since the changes made are under the cloak of Europeanization and the Westernization of this unruly region. What both periods, pre- and post-Milosevic, showcase is the relation of (il)legality with lack of public transparency.

Weiss' attempt to suggest that architecture during the SFRY was largely homogeneous in style is odd. Regional distinctness was pursued by architects and domestic owner-occupiers alike. The 1961 economic reforms were driven by the opening of Yugoslav borders and the legalization of emigration; the strategy was to keep the economy afloat with the remittances of Yugoslav emigrants who worked abroad, while also easing the rising national unemployment. Those who worked in Germany and Austria would, with their hard-earned money, often build large and lavish residential dwellings that appropriated traditional Germanic and Austrian styles. These influences were more apparent and style more dutifully followed in areas such as Croatia.

In areas such as Serbia, the appropriations were more opulent, blending the styles inherited from the Turks during the Ottoman Empire with the traditional Germanic or Austrian styles. Just how much hybridization and influence were taken from different architectural styles and cultural traditions depended on the personal aesthetics of the owner of the dwelling. When structures were designed by architects, they became no less distinctive. These residences also accommodated businesses ranging from hairdressing and beauty studios to dentist's surgeries; hybridity was achieved through lack of programme distinctness. The scale and mixing of architectural programmes perhaps anticipated Rem Koolhaas' 1995 thesis that classification of architecture needs to be based solely on scale, not on things like building typology, or location or construction techniques.[57]

With the 1948 split with Stalin's Russia, Yugoslavia also moved away from centralized socialist planning. Balkanization, that is, decentralization, was integrated purposely into the construction of mass housing after WWII. Architectural differentiation was clearly a marker of balkanization, despite this balkanization in Yugoslavia also being politically synonymous with socialism, anti-fascism and bringing together different ethnicities. For example, the interpretation of double-tract units in mass housing in Serbia was two units that varied in function and shape connected vertically into a unitary structure.[58] In other republics such as Croatia, the interpretation was a corridor joining two functional zones.[59] The difference of interpretation in mass housing also led to varying façade treatments; from Serbia's interest in experimental patterns and pulling out volumes to Croatia's interest in colour, to Slovenia and Bosnia-Herzegovina's focus on appropriating the dominant Western aesthetic of modernism.[60] The architecture of balkanization was not a question of uniform mass production, but a complex operational procedure whereby function was also dependent on experimentation in pre-fabrication, volumes and façades. What brought the diversification together is economical affordance and speed of construction; balkanization as implemented in the former Yugoslav context being an event of transgression.

Balkanist transgressions and the Mikser House

The radical potential of Balkanism (especially in the Yugoslav case of balkanization, in which territories where different South Slav peoples lived were internally integrated, meaning that the entity thus created was heterogeneous) is that it demonstrates alternative ways a society can be set up through events, leisure and everyday life as well as rethinking the preoccupation with land and power. In other words, the potential is found in the frictional and deviant Balkanism, comparable to Deleuze's description of nomadism, since for Deleuze "the nomads are those who don't move on, they become nomads because they refuse to disappear."[61] It is the refusal to disappear that is the event of transgression. Thinking *event* in relation to Balkanism is to think how mapping events in terms of the political landscape in which they arise proffers thinking event as that which eclipses the 'self-evident' understanding of that event in terms of its originary articulation. In other words, the focus is more on how the event articulates power, the effects it has and how these may be used as an operative tool to (de)stabilize and challenge the dominant understanding of history of this region.

One contemporary example was seen in the Don't Drown Belgrade/We are not Handing Over Belgrade political campaign discussed earlier. Another has been around since 2008, and it utilizes Balkanism in a very different way. The independent design and cultural centre Mikser House and the annual Mikser Festival that started in that year – the largest festival of innovation and creativity in the region – skirt local and global politics in an attempt to create a multidisciplinary platform where the distinctions between architecture, media design, spatial activism, education and economics start to falter. This platform reveals an operation of Balkanism beyond the conventional rhetoric of barbarism.

Mikser House was cofounded in 2009 by a husband-and-wife duo, Ivan Lalic (a playwright) and Maja Lalic, an architect who studied in Belgrade and New York. Her return from the US was prompted by the termination of Milosevic's rule in 2000 and the promised surge of democratization in the country. Initially located in Savamala, Mikser House was part of a larger desire to revitalize the area along Belgrade's Sava, the zone currently being redeveloped as part of Belgrade Waterfront. Other notable cultural centres that emerged within Savamala were KC Grad and the Spanish House as well as a number of alternative cafés and bars. The criticism that has dogged these initiatives has been that the touted cultural revitalization is no more than exploitation of simulacra in a drive to pile high the symbolism that goes with anything that is understood as arts and culture, even where it is contingent on capital accumulation and the removal of the less privileged layer of society.[62]

If the criticism is valid, this amassing of the symbols of culture was nevertheless not spared by the Belgrade Waterfront plan. In May 2017, Belgrade's Mikser

House had to leave its premises, once a vacant mechanical workshop, and temporarily close its Belgrade branch. The Belgrade Waterfront 2012 master plan had marked the area of the Mikser House as a prime zone for "urban regeneration," and this was now to be put into effect. Two things set the Mikser House apart from other design and cultural centres in this Balkan region. The first was its yearly thematic Mikser Festival.[63] In 2017, the Festival operated from Dorcol (an industrial zone in proximity to the port, one occupied by an eclectic population from a low socio-economic background and whose post-1999 redevelopment is at a halt). The second was seen in the plan to expand its activities into other Balkan and former Yugoslav areas, with the intent of showcasing Balkan creativity and innovation through coexistence and beyond the Balkanist derogatory rhetoric. The first such expansion, albeit short-lived, was the opening of Mikser House in Sarajevo in September 2017.[64] In many ways, the ethos of Mikser House reinterpreted the 20th Century Yugoslav ideology by applying balkanization in reverse to the norm. Rather than fragmenting areas and isolating different ethnicities, Yugoslav balkanization was associated with multiethnicities and constituent nations coexisting in the same territory.

What the independent Mikser House and Mikser Festival showcase is a belief in change, and a more tolerant and heterogeneous world. In many ways, the event is that the very existence of belief prompts transgression of control, since for Gilles Deleuze

> [i]f you believe in the world you precipitate events, however inconspicuous, that elude control [. . .] Our ability to resist control, or our submission to it, has to be assessed at the level of every move. We need both creativity and a people.[65]

It is true that Mikser's instatement of balkanization is tied to consumerism and advertising; if Mikser House follows the originary idea of expansion, it is bound to become more mainstream and at the same time driven by the desire to increase profits. However, what has remained constant in the initial conception and during its period of expansion (exemplified by the short-lived existence of a Mikser House in Sarajevo) is the desire to create a space that can showcase art, design and cultural knowledge and practices specific to the Balkans. Both the annual festival and the Mikser House are driven by an ethical agenda of eliminating racial, ethnic, religious, gender and age boundaries through workshops, debates, youth programmes, exhibitions, lectures, dance and cinema.

Mikser House does not discredit consumer culture but uses it to attempt to shift the thinking around Balkanism; it shows an alternative thinking when it comes to economics and its relation to solidarity among regions that only very recently, in the 1990s, fought over territory. What Mikser House also does, despite being profit-oriented, is to showcase, through a multiplatform and multiethnic approach found in arts and culture, the presence of a long-standing history of centuries-old socio-political cooperation in this Balkanist region. That this platform is also driven by a belief and a desire to bring together kids from across the former Yugoslavia around music and performance, as seen during the opening of Sarajevo's Mikser House,

Figure 6.11 The façade of the no-longer-operating Mikser House in Sarajevo. Parts of the façade read, "This land is for all. This is a house from a dream."

is a step towards building friendship and tolerance. This is noteworthy considering that current education in each of the former Yugoslav republics is tainted by efforts to short-circuit history with teaching materials that negate solidarity. From this perspective, the Mikser House and Mikser Festival reflect Bernard Tschumi's thinking whereby the event is a different mode of thinking space.[66] This position somewhat parallels Rem Koolhaas' thinking when during his visit to Belgrade in 2003[67] he advocated the importance of Belgrade capitalizing on its potential to lower urban standards, rather than compete with standards of living found in Western Europe, and thus always remaining a B or C version.[68] The Mikser House and Mikser Festival have appropriated his conceptual position. Where Koolhaas, and to an extent Mikser House, undermine the potential of Belgrade, Balkans and Balkanism is that they associate the city, the region and the discourse with a set of measurable values. As such, the radicality of thinking *event* in relation to Balkanism is weakened since Balkanism is trapped by being dependent upon the Western ordering of high-low values and imposition of Western power structures.

However, the opening of a centre for refugees – Refugee Aid Miksaliste – with the first wave of arrivals from the Middle East, and as an extension of Mikser House, showcases the attempts to extend solidarity. The overall ethos is oriented towards associating the Balkans and Balkanism with vitality and creativity, rather than with violence and conflict. More recently, and as a result of space infringement – as seen with the closure of the Mikser House in Savamala – the Mikser House has also been turned into an educational caravan that travels to different cities of Serbia addressing issues from urban planning to civil society. The workshops bring together various experts in the fields as well as NGOs to think ways of becoming more proactive and participatory in the current political climate.

Thus, thinking events via Balkanism and balkanization are dependent upon mobility and active participation which seeks to both challenge and transgress the transitionary impositions of control in the form of delineation of values and lack of transparency to do with urban development as well as commercialization and privatization. The potential lies in Balkanism being seen as a discourse of indefiniteness, the Balkans as a region where "inhabitants do not care to conform to the standards of behavior devised as normative,"[69] and that balkanization as fragmentation can be reversed, as seen during the formation of the 20th Century Yugoslavia. What Balkanism and the Yugoslav version of balkanization demonstrate is not only that an outside exists, but also that the discourse may be used to re-think the political meaning that the society can become more open and constituent. After all, Rebecca West, in her 1942 *Black Lamb and Grey Falcon: A Journey Through Yugoslavia*, described this territory as "a decentred border space, which deconstructs linearity."[70] Likewise, not only does the centre not exist, but also neither does the conception of space as a stable entity with set origins.

Notes

1 Weiss, "National, Un-National," *Nationalities Papers: The Journal of Nationalism and Ethnicity*," 104.
2 Rastko Močnik, "Social Change in the Balkans," *Eurozine*, 20 March 2003, <www.eurozine.com/social-change-in-the-balkans/>
3 Lorenc Agerman, Eduard Frojdam and Djan Gildji, *Beograd Gazela: Vodic kroz Sirotinjsko Naselje* (Beograd: Rende, 2009), 20.
4 Agerman, Frojdam and Gildji, *Beograd Gazela*, 20.
5 Agerman, Frojdam and Gildji, *Beograd Gazela*, 42.
6 Bojana Barlovac, "Reconstruction of Main Belgrade Bridge Kicks Off," *BalkanInsight.com*, 1 July 2010, <www.balkaninsight.com/en/main/news/29160/> Note: this bridge was not targeted in 1999.
7 Barlovac, "Reconstruction of Main Belgrade Bridge Kicks Off."
8 Zlata Vuksanović-Macura, *Zivot na Ivici* (Beograd: Orion Art, 2012).
9 During NATO's 1999 campaign, Branko's Bridge was utilized as a platform to expose and challenge the national and international political inconsistencies of NATO's targeting of Belgrade. The people became human shields to protect the bridge from being targeted. This is despite the attempt to paralyze all movement and protests taking place in Belgrade's urban spaces. Arguably, the Romani below the Gazela Bridge also shielded that bridge from destruction. During the 1999 Operation Allied Force, NATO destroyed 24 road and railway bridges and overpasses and damaged another 36. The destruction of some of these transport routes is peculiar, considering that a number were located in central and northern Serbia, areas which are not in direct proximity to Kosovo and not of obvious military significance. For example, three bridges were destroyed/damaged in Novi Sad, the capital of Vojvodina, Serbia's northern province.
10 Amnesty International, "Roma Families in Belgrade Face Forced Eviction by Government Agency," 2 November 2011, <www.amnesty.org/en/news/roma-families-belgrade-face-forced-eviction-government-agency-2011-11-02>.

11 Amnesty International, "Roma Families in Belgrade Face Forced Eviction by Government Agency."
12 Amnesty International, "Roma Families in Belgrade Face Forced Eviction by Government Agency."
13 The 1937 complex included Italian, Hungarian, Romanian and Czechoslovak pavilions were also built. A Turkish pavilion was added in 1938 and a German one in 1939.
14 Forensic Architecture and Grupa Spomenik, "Living Death Camps," in *Forensis: The Architecture of Public Truth*, a project by Forensic Architecture, Centre for Research Architecture, Department of Visual Cultures, Goldsmiths, University of London (Berlin: Sternberg Press, 2014), 198.
15 Forensic Architecture and Grupa Spomenik, "Living Death Camps," 191.
16 The first urban plan for New Belgrade, called 'Singidunum Novissima,' was prepared by architects Rudolf Perco, Erwin Ilz and Erwin Böck in 1922. This plan proposed boulevards and neo-baroque architecture. In 1923, the architect Djordje Kovaljevski proposed an expansion of Belgrade's Old City to incorporate the current day areas of New Belgrade and Zemun. Another notable urban plan for New Belgrade was prepared by the architect Jan Dubovy in 1924. In Dubovy's plan, significance was attached to the value of fertile land, as Dubovy saw agriculture as a foundation for a healthy city. Agricultural labour, according to Dubovy, solved larger social and economic questions. The architect Milorad Pantovic's 1940 plan was prepared as part of a tourist exhibition at the Fairground Pavilion built in 1937 in New Belgrade (the Fairground, as will be discussed, was built despite no urban plan having been approved). It was not until 1941 that the architect Dragisa Brasovan proposed an urban plan where New Belgrade was broadly imagined as a central living and administration hub for Belgrade. His plan dealt with more complex questions such as culture as well as pragmatic ones such as transport. Since the plan was devised in 1941 at a time when Belgrade was being bombed and occupied by the Nazis, this plan was disregarded in the post-WWII development due to its associations with Fascist rule. This is despite Brasovan being neither a Nazi or a supporter of the German invasion, nor his proposal explicitly reflecting a particular political ideology. For more information, refer to Ljiljana Blagojevic, *Novi Beograd: Osporeni Modernizam* (Beograd: Zavod za udzbenike, Zavod za zastitu spomenika kulture grada, Arhitektonski fakultet, 2007).
17 Zaklina Gligorijevic, "Can City Development and Identity Grow in Harmony?" *42nd ISoCaRP Congress 2006.*
18 Eva Vanista Lazarevic, Boris Koruznjak and Mirjana Devetakovic, "Culture Design-Led Regeneration as a Tool Used to Regenerate Deprived Areas. Belgrade – The Savamala Quarter; Reflections on an Unplanned Cultural Zone," *Energy and Buildings* 115 (1 March 2016): 6.
19 Lazarevic, Koruznjak and Devetakovic, "Culture Design-Led Regeneration as a Tool," 6.
20 Ljiljana Blagojević, "Urban Regularisation of Belgrade, 1867: Trace vs. Erasure," *SAJ* 1 (2009): 28–30.
21 Blagojević, "Urban Regularisation of Belgrade, 1867," 34–35.
22 Blagojević, "Urban Regularisation of Belgrade, 1867," 34–35.
23 Zygmunt Bauman, *In Search of Politics* (Stanford: Stanford University Press, 1999), 5.
24 Bauman, *In Search of Politics*, 5.
25 Jürgen Krusche. "Bureau Savamala Belgrade," in *Bureau Savamala Belgrade: Urban Research and Practice in a Fast-Changing Neighborhood*, ed. Jürgen Krusche and Philipp Klaus (Berlin: JOVIS, 2015), 15.
26 Ksenija Lalović, Uroš Radosavljević and Zoran Đukanović, "Reframing Public Interest in the Implementation of Large Urban Projects in Serbia: The Case of Belgrade Waterfront Project," *Architecture and Civil Engineering* 13:1 (2015): 43.
27 Lalović, Radosavljević and Đukanović, "Reframing Public Interest," 43.

28 Some of these ideas have been addressed in respect to some American cities such as Los Angeles. See Mike Davis, *City of Quartz* (London: Verso, 2006); and David Harvey, *Justice, Nature and the Geography of Difference* (Hoboken: Wiley-Blackwell, 1996).
29 Andrew Ballantyne, "Architecture as Evidence," in *Rethinking Architectural Historiography*, eds. Dana Arnold et al. (London: Routledge, 2006), 48.
30 Robert Venturi et al., *Learning from Las Vegas: The Forgotten Symbolism of Architectural Form* (Cambridge, MA: MIT Press, 1977), 34–35.
31 In Serbian, the linguistic use and interpretation of the word 'duck' accommodates the meaning to be also associated with fraud.
32 The president of the party is Aleksandar Vucic, who is also the President of Serbia.
33 The other part of the waterfront development is the side that faces the Danube, the Port of Belgrade. The port, which is located between the city neighbourhood of Dorcol (an industry zone and an eclectic area occupied by the population of a low socio-economic background), and the river Danube, prospectively occupies one of the most prestigious pieces of land in Belgrade. The privatized Port of Belgrade and its development company initially engaged Daniel Libeskind to create a master plan for the Port area, with the presentation of this design having taken place in March 2009. The master plan proposed a luxury waterfront plaza with a commercial/residential complex united by twin towers, suggestive of the two rivers that pass through Belgrade (Danube and Sava). While Libeskind's proposal was not implemented, neither has anyone's else. The project is largely at a halt.
34 *Buvljak* opened in 1994 during the disintegration of the SFRY. During the war, the market sold goods that were often (illegally) transported from countries such as Bulgaria, Hungary, Romania or even China. *Buvljak* is still in operation today, serving a significant layer of Belgrade's population that is in need to survive. The market will be moved to a different location in New Belgrade in the near future.
35 Brutalism arose as an architectural style in the 1950s, and is known for its exposed concrete, monolithic and block-style structures.
36 Miloš R. Perović, *Iskustva Prošlosti* (Beograd: Zavod za planiranje razvoja grada Beograda, 1985), 126.
37 Urban modernization in the former SFRY, as seen with the construction of New Belgrade as well as New Zagreb, was not completed; numerous blocks remained vacant.
38 Sonia Hirt and Mina Petrovic, "The Gates of Belgrade: Safety, Privacy, and New Housing Patterns in the Post-Communist City," *Problems of Post-Communism* 57:5 (2010): 6, accessed May 12, 2011, <www.spia.vt.edu/SPIA/docs/shirt/The_Gates_of_Belgrade.pdf>.
39 Ljubica Gajević, "New Belgrade Urban Fortunes: Ideology and Practice Under the Patronage of State and Market" (Masters Dissertation, University of Catalonia, Barcelona, 2011), 97–98.
40 The Museum of Contemporary Art, partially destroyed by the 1999 bombings, was reopened at the end of 2017. The 18-year delay in fully repairing the Museum, which was initially designed to exhibit predominantly regional art, is suggestive of the lack of value placed on Balkan and Yugoslav art.
41 Neil Smith, "New Globalism, New Urbanism: Gentrification as Global Urban Strategy," *Antipode* 34:3 (July 2002): 427–450.
42 Michael Hardt and Antonio Negri, *Multitude: War and Democracy in the Age of Empire* (London: Penguin Books, 2004), 21.
43 Gilles Deleuze and Felix Guattari, *Anti-Oedipus: Capitalism and Schizophrenia* (New York: Penguin Books, 2009), 500.
44 It must be noted that the paradox of Yugoslav instatement of self-management – associated with decentralization of power and affordance of individual rights through the establishment of workers councils – is that it was still dependent on the hierarchical structure of the socialist party, and implicitly the governing imparted through the 27-year long presidential reign of Tito (1953–80). He was also the Prime Minister of Yugoslavia from 1944 to 1963.

45 Jean-François Lyotard, *The Assassination of Experience by Painting, Monory* (London: Black Dog, 1998).
46 Michel Foucault, *Discipline and Punish: The Birth of the Prison* (New York: Pantheon Books, 1977), 36.
47 Kolstø, "Western Balkans' as the New Balkans," 1250.
48 Weiss, "National, Un-National," 104.
49 Kai Vockler, " 'Balkanology' and the Future of the European City," in *Balkanology – SAM No. 6*, edited by SAM, Francesca Ferguson and Kai Vockler (Basel: Christoph Merian Verlag, 2008), 11.
50 Bakic-Hayden and Hayden, "Orientalist Variations on the Theme 'Balkans,'" 8; and, Bakic-Hayden, "Nesting Orientalisms," 920.
51 Grubačić, *Don't Mourn, Balkanize!* 42–156.
52 Srdjan Jovanović Weiss, "School of Missing Studies," in *Did Someone Say Participate? An Atlas of Spatial Practice*, ed. Markus Miessen and Shumon Basar (Cambridge, MA: MIT Press, 2006), 241–257.
53 Srdjan Jovanović Weiss, "Milosevic as Architect," in *Networked Cultures: Parallel Architectures and the Politics of Space*, ed. Helge Mooshammer and Peter Mörtenböck (Rotterdam: NAi Publishers, 2008).
54 Weiss, *Almost Architecture*.
55 Uroš Čvoro, *Turbo Folk and National Identity in Former Yugoslavia* (London: Ashgate, 2014).
56 Weiss, *Almost Architecture*, 39.
57 Rem Koolhaas, *Bruce Mau and OMA, S, M, L, XL* (New York: Monacelli Press, 1995).
58 Đorđe Alfirević and Sanja Simonović Alfirević, "Urban Housing Experiments in Yugoslavia 1948–1970," *Spatium* 34 (December 2015): 1.
59 Alfirević and Alfirević, "Urban Housing Experiments in Yugoslavia 1948–1970."
60 Alfirević and Alfirević, "Urban Housing Experiments in Yugoslavia 1948–1970," 1.
61 Gilles Deleuze, *Negotiations, 1972–1990* (New York: Columbia University Press, 1995), 138.
62 Ana Vilenica, "The Art of New Class Geography of the City: Culture-Guided Urban Regeneration Serving the Modernization of the Periphery," in *The Grey Zones of Creativity and Capital*, ed. Gordana Nikolić and Šefik Tatlić (Amsterdam: Institute of Networked Cultures, 2015), 59.
63 The theme for the 2017 Mikser Festival was 'Migrations.' Other themes ranged from 'Sensitive Society,' 'Itch,' 'Utopia' and 'Sustainable Utopia' to 'Urban Transformations.' The 2019 theme is 'Circulate!' and is oriented towards the future and the phenomena of life and work resilience.
64 The Sarajevo project was impressive in its ideas and actual physical development. However, the project's dependence in large part on loans, which could not be paid within negotiated periods, led to its demise. The lesson in this was that an alternative needs to be started with caution, and certainly not on the basis of dependence on economic loans.
65 Deleuze, *Negotiations*, 176.
66 Bernard Tschumi, *Architecture Concepts: Red Is Not a Color* (New York: Rizzoli International Publications, 2012), 177.
67 Rem Koolhaas' visit was made possible by Mikser House and the Serbian Ministry of Culture.
68 Weiss, "National, Un-National," 97.
69 Todorova, *Imagining the Balkans*, 3.
70 Rebecca West, *Black Lamb and Grey Falcon: A Journey Through Yugoslavia* (Edinburgh: Canongate Books, 2006), 187.

7 Present spaces, present times

Other spaces, other times

Present

Political tensions of Balkanism, balkanization and architecture

With the geopolitical zone of the Balkans being historically complex, the countries that now form the political zone of Western Balkans (which includes Serbia) are those whose physical and psychological spaces (if not their geographical ones) are *still* perceived as an abject inconvenience and an anomaly in the privileged Anglo-American and Western system. Indeed, the necessity to eliminate this type of space is given paramount attention; the drive to absolve this inconvenient space is made possible with its association with barbarism and perennial violence. In terms of the former Yugoslavia, the means of achieving this has been a multi-faceted and complex process that has seen the country's territory first balkanized and then destroyed in space and time, with consequent effects on the urban and architectural spaces of everyday life.

The balkanization that was implemented with the 1990s disintegration of Yugoslavia into separate enclaves was also a means of placing restraints on the exercise of a diversity of beliefs, values and relations; it was indicative of elimination of coexistence. Simultaneously, its enactment prompts us to rethink our current value systems and conventions to do with defining spatial violence and human agency; the meaning and deployment of both grow less clear, particularly because the violence perpetrated in the 21st Century is understood through the lens of the 20th Century.

Studying the complexity and the transitionally undecided state of former Yugoslavia reveals distinctive ways in which balkanization is being deployed; the understanding of the intents is the understanding of the contemporary implementation of balkanization. The focus on former Yugoslavia in general, and Belgrade in particular, is largely because any imposition of power for purposes of evolution and attaining civility becomes more obvious as the power structures implemented are coming into contact with a zone of counterapproach to balkanization by affirming and expanding heterogeneity. The study of balkanizing processes past and present reveals the political tensions inherent in architecture and urbanism, including the fragility and constraints of power and its structural logic. In other words, the balkanization of cities and architecture in the aftermath of the 1990s dissolution has been an opportunity to reduce the depth and complexity of history, both that of the immediate past and of more distant times. Matters to do with identity were also brought into being and implemented symbolically or pragmatically; predominantly for purposes of classifying violence in the name of those who may speak, that is, those who do not have a right to voice.

The ideological and symbolic aspects of architecture were often narrated through mass media, which portrayed the 1990s dissolution largely as a result of ethnonationalism. However, the balkanization of Yugoslavia was more tied to eliminating its association with communism, socialism and anti-fascism by reducing the complexity of history through urban and architectural spaces, including the breadth of violence taking place on the ground. Balkanization in this instance was deployed for purposes of categorizing and managing data and information, to be processed and transmitted to the public to perpetuate a uniform message that Western and Anglo-American intervention was undertaken to bring civility and peace. It was also done to map the flows of people and commodities and extend military control into every aspect of life by blurring the line between the military and the civilian. Two examples from the kind of peace brought to Kosovo in the aftermath of 1999 were the apartheid created for Serbs and Romani and the piece of territory that was sequestered for a US military base. Meanwhile, in FB-H, enclaves were legally implemented through the 1990s Dayton Agreement.

Where the potential of Balkanism lies is in its liminality. This was seen in the public protests that occurred in Belgrade during the NATO bombings; the 'abnormality' was extended and space reinterpreted despite Western warnings and threats. Balkanization may also be used in opportunistic ways, as symbolically suggested in Kusturica's films, where new territories can be created again for purposes of ambiguity and in the name of joy and diversity. The course of history can be altered, both symbolically and pragmatically; the Generalstab complex symbolic of the real Sutjeska Offensive that brought into existence the affirmative balkanization through extension of difference and heterogeneity of the second Yugoslavia as well as anti-fascism.

The politics of identity is a significant aspect of balkanization and architecture. The unmaking and making of cities and space tends to be a complex interplay of historical short-circuiting and fabrication of myth. This was seen in the reconstruction of Croatia, FB-H and Serbia. Thus, to understand the complexities of identity reconstruction, there is a need to engage with the politics of rubble. Reconstruction extends beyond physical remaking, as seen with the rebuilding of the Old Bridge in Mostar; the reconstruction often making apparent that which is not ready to come together. Having said this, any form of balkanization including political matters to do with the identity of architecture needs to be approached from the perspective of more than one history. Otherwise, the framing and engagement with history remains polarized; violence required to be visible and seen as it was perpetrated by Others, and that which needs to remain invisible and not seen as it was enacted in the name of civilization and humanitarian values thus allowing the cleansing of that violence. Moreover, when balkanization is implemented for purposes of policy control and in the name of defence, as seen with the reconstruction of the Generalstab complex, there is less possibility to penetrate the power vacuum of perpetual present. Likewise, it is an opportunity to control the access and use of data and information on the basis of security in that security is framed as an issue of defence rather than as a social and developmental concern.

From a pragmatic perspective, NATO's intervention was geared towards the introduction of the market economy, as seen with the Rambouillet Agreement. Balkanization was enacted for purposes of attaining civilization and instrumentalized through urban infrastructure via privatization and control of entities from construction firms, power plants and telecommunications to banks and matters to do with shipping. This is the politics of greed. Infrastructure has also been used for purposes of eliminating any sites that breach the norm, such as the Romani settlements beneath Belgrade's Gazela Bridge as well as from the city centre of Belgrade. This is an example of domesticated xenophobia in Serbia, and internationally towards the Romani. The sites of violence whose borders are less visible are those created by NATO's use of DU; that violence of balkanization is silently enacted years after the 1999 intervention, since DU is limited neither by time nor by territorial boundaries.

More broadly, Western values have been attained together with the removal of social welfare and rights to public good. From a socioeconomic and ethnic perspective, the alterities of Balkanism and the Yugoslavian type of balkanization are still possible. The work by Mikser House, Mikser Festival and Refugee Aid Miksaliste is very much geared towards solidarity with a multiplatform and multiethnic approach. The activism of Don't Drown Belgrade/We are not Handing Over Belgrade in response to the Belgrade Waterfront development showcases that events of transgression and change are still possible. From an architectural and urban perspective, the implementation of balkanization in Yugoslavia showcased that mass production, social housing and profit are possible without letting go of creativity and experimentation. Moreover, globalization was approached in a way to think solidarity, affordance and heterogeneity rather than the current practice of discord, polarization and homogeneity.

Global politics, balkanization and the camp

Balkanization extends beyond the Balkans. The UK's decision to leave the EU is a sign of not only balkanization, but also the beginnings of regional fragmentation of many other European countries. North Italy's secessionist tendencies are on the horizon and, as in Spain and Yugoslavia, are driven by reluctance to share economic wealth in a way that subsidizes neighbours who are socioeconomically less fortunate. While autonomy may seem to promise independence and economic gain, the mid- to long-term effects are potentially troublesome since such newly balkanized regions find themselves more easily dominated by global powers such as NATO. While Yugoslavia's 1990s balkanization may be seen as a different scenario due to the country's Titoist and, therefore, not democratic dispensation, the very premise of balkanization as a method and a system of analysis intersecting with fields that extend beyond the borders of the Western Balkans – such as sociology, the law or the urban – suggests that fragmentation is pursued for purposes of administration and control, as well as to place restrictions on diversity through xenophobia and scapegoating.

The global world of today is not only remade and reorganized as a result of ongoing ways of destroying and militarizing the urban by destroying public space through privatization, militarization of security and implementation of a permanent sense of fear, as well as using the city as a space to enact warfare.[1] It is also a fortified camp of aestheticized threat and imaginary fear. The understanding of the constructs embedded within the rhetoric of risk would undermine the need for permanent security and the construction of borders (real or virtual) to prevent coexistence with the Other. Understandably, this is not in the interest of the military since fear is used as a tactic and as that which both provokes and maintains the presence of risk. It is a cyclical process; to make defence and security a matter of priority, fear needs to be invoked regardless of how real or imagined the threat may be. Fear, in this instance, is used as a tool to isolate and to legally toughen borders; it is the agent and the motive behind every implementation of economic and defence measures. It is used to obscure the fact that we are living in a time when it is becoming ever harder to step back and recognize that such imaginary is governed.

Control is found at the intersection of imagery and forced perspectives used for purposes of convenience and entertainment. An ability to decipher the aesthetics and simulacra of imagery would make it possible to step out of the infantile, mechanistic and purely visual mass media industry in which we engage in. However, unlike Giorgio Agamben's thinking on the camp, in which "there was no space for rest, reflection and comfort: work, finding something to eat and survival were parts of a daily battle, which meant that the prisoners were in permanent movement,"[2] mobility in the current globally connected society is constrained by the need to be physically inert due to imminent threat. The ability to perceive and make sense of the 'threat' depends on its presentation in the mass media. Though, this reality assumes short-term memory loss on a scale that forgets even events in the immediate past; moreover, a future beyond the one projected by the dominant political structures is hard to imagine since the political realm is activated and stimulated around a recurrent circulation of one-liner narratives and representations. It is this short-term memory vacuum that produces a situation of not seeing or believing in alternatives other than those which currently exist, which creates a fertile ground for the present to become even more rigid and extreme in the name of supposed defence and security.

Contrasting Agamben's conception about the camp, the current camp is always altering its formation precisely because it is dependent on how the Other is framed. That framing is contingent on whether the Other's system manifests political oppositions, and/or alternatives, to the privileged political structures in Anglo-America and Western Europe, or if those Other zones are willing to subserviently accept the imposition of Western values. Those who challenge the systems and conditions presented are silenced through association with violence. Thus, when the derogatory Balkanist agenda is deployed, it is to deny the possibility for alternative values and relations, that is, to eliminate spaces that operate outside Western Enlightened values. Paradoxically, these Western values enacted on behalf of humanitarianism, human agency and justice are compatible with military interventions. The values of civilization have arisen on foundations of violence and wars. Likewise, current moralization is established by the ethos of NATO, the new superpower. With the wars in Bosnia-Herzegovina and Kosovo, NATO has established itself as an organization that drives peace-building and facilitates processes for civilization in southeast Europe.[3]

Global politics is nothing other than violence deployed as a mode of military operations and transitional politics, despite this violence being purified as it is decreed in the name of human agency and civil values. The calamity of such replicatory violence is even greater when Western Balkans countries agree to partake in binding dialogues with Western Europe and Anglo-America that are constructed on the foundations of the derogatory Balkanist imaginary. The dialogue is not even-handed; instead, it takes the side of and privileges one type of violence and indirectly misrepresents another type of violence. In terms of the Balkans, misrepresentation is achieved through dehumanization. Violence is condoned precisely because this part of the world is said not to have reached the level of maturity, found in the rest of Europe and Anglo-America, that would sustain a functioning multiethnic society. To comprehend the Balkans, after all, requires grappling with complex and interrelated histories of cultures, beliefs, practices and values. It is a task that confounds the conventional cognitive map of understanding spaces and events via fixed coordinates or the peculiarly European nation state formation through homogenization. It also supplies the opportunity to mark this Other space as an anomaly and a black hole of barbarity. Such sidestepping of the norm is deemed to include the inability to engage in civility by virtue of the fact that it challenges that very norm.

Humanism of normality

The norm is not just a social construct implemented for purposes of disciplining and compartmentalizing humans, but also one of articulating pressing concerns. This is particularly relevant when it comes to questions of displacement, that is, human resettlement, regardless of whether it is generated by conflicts, natural disasters or gentrification. With the statistics indicating that 68.5 million people have been forcefully displaced globally, what is even more concerning is that since the early 1990s close to 35,000 people were found to have died while attempting to cross borders to get into Europe, or in detention camps. When it comes to Europe, these borders are located along the territorial line of the Western Balkans and the Mediterranean (southern Europe).[4] The need of refugees to cross the border into the EU while not accepting assistance from any of the Western Balkans countries is driven by Europe's decision that accepting humanitarian aid is tantamount to expressing willingness to remain living in countries such as Serbia. Remaining in the Western Balkans disqualifies the refugee from seeking aid from more economically stable countries.

The question of humanitarian aid is a matter not just of Foucault's bio power but also is concerned with difference in that difference is to be devalued and borders need to be policed against it. This was seen in the 23 September 2015 speech given by Donald Tusk, the president of the European Council, who advocated the need to reinforce border control as a matter of "responsibility and common sense," lest Europe experience a "political catastrophe" from allowing the refugees to settle.[5] Refugees are thus synonymous with threat and are directly agents that will erode peace, freedom and the identity of the EU. This is freedom, but freedom *from* difference.

To have freedom *in* difference would mean the elimination of obstacles to thinking and acting differently from the current political reality. Having this freedom is to simultaneously have the capability to critically analyze the legitimation and extension of control and to think beyond the current implementation of violence. Perhaps this type of freedom is discouraged since the established culture is civil only if border fortifications and systems of control are erected and maintained. Unlike the medieval forts where the walls were visible, current forts are softer and more 'democratic;' they are instated through legal policies and the rhetoric of threat. Even though the refugees are fleeing persecution, their existence is conflated with the language of barbarism and threats to Western security. To invoke security is to allow the possibility to limit civil liberties – a technique made apparent after 9/11 when US authorities 'rightfully' subjected migrants to interrogation based purely on their appearance and a supposition that they would perpetrate violence should they be granted entry. The invocation of security is an attempt not only to manage uncertainty but also to establish uncertainty as a persistent image associated with peril. It is to project an image of the present as well as the future before it even occurs. Heidegger's thinking to do with all beings living their life in the shadow of death[6] is now being overtaken by the situation of all beings living their life in the shadow of fear. Fear is the seat of power over life, and the implicit destruction of life.

The humanism of normality is the politics of discrediting alternative futures. To believe in the necessity of enforcing border security (physical, legal and socioeconomic) measures is to find meaning in protection *from* difference; implemented through finding justification and meaning in violence inflicted upon the Other. With the alleged need to protect against threat, the necessity for external and internal, as well as general and particular, security also grows in partnership with the rising fear. The great danger is of fostering a belief in the enduring righteousness of these systems and that they should continue indefinitely. A greater danger, though, is the belief that there is no outside. Hardt notes that the corollary of Foucault's thinking that power "comes from everywhere, [and] that there is no outside to power, [is that] he is also denying the analytical separation of political society from civil society."[7] In other words, there is no possibility to resist or to create an alternative reality since the operational mode of power is circular for Foucault. There is no outside to the history prescribed.

Balkanization implemented in the name of 'humanism' itself is a new way of imposing Western values and diplomacy that allows the use of force. It is a tale of morality narrated in terms of obedience and submission. However, since the intentions are 'just,' the violence is automatically absolved. These values enacted in the name of human rights, justice and freedom have been made possible precisely by displacing what Bauman identifies as the "*truth* of modernity" in that this identification facilitates the ability to selectively enact and display these values.[8] Kriss Ravetto elaborates upon Bauman's thinking by saying that such selectivity has afforded the West the chance to associate "less modern or civilized figures and nations [. . . with. . .] Saddam Hussein, Stalin, Serbia, and Cambodia."[9] The upholding of these values has been rendered possible first by isolating particular areas and second by using that isolation as an opportunity to construct knowledge to do with lack of civility found in those territories and/or valued amongst their people. In contemporary times, the construct gains further credibility from being posed in terms of economics, security and defence of civil values and human rights.

The belief in the existence of Western modernity's justice and freedom is the belief in violence since the formation of modern nation states was contingent on balkanization where newly created states were largely ethnically homogeneous. It would seem that the "*truth* of modernity" is the veiling of violence and bloodshed; the invocation of liberty, fraternity and humanity during the French Revolution coincided with mass killing. Also, the very premise of humanity is anything but humane since it is moralized on the basis of classifying people and places in terms of practices, convictions and imaginaries; the contemporary practice of humanity not only fabricates history but also prevents it from existing.

Despite, or because of, the heritage of values from the perspective of the "*truth* of modernity," it was necessary to identify areas such as the Balkans as zones of historical barbarity where the possibility of humane and civilized ways like those found in Europe and the West did not exist. This belief in a lack of moral values and humanity in the Balkans is not only paradoxical, considering the Western history of colonialism and balkanization, but also unproductive, as it traps Western history in an imaginary construct of heroism and progress. That this progress is flawed is made evident by the fact that, despite significant technological advances, the hand of solidarity is still not extended unless it comes with the imposition of values geared towards the market economy. The imposition of values is also dependent on eliminating coexistence with difference, unless that difference is kept outside designated borders; legal regulations and social values are built upon delimiting difference.

To participate in the imagery of modernity's values is to accept that certain liberties, such as freedom to think and live outside the prescribed norms, can be suspended if necessary, for purposes of defence. The rhetoric of security not only makes violence more palatable when distanced from Western borders, but also requires defence from it in that the media screen has been able to portray the need to guard the borders to ward off violence. Blocking violence is simultaneously contingent upon suppressing the efforts of mobilizing thought if that thought challenges assumptions about the vilified Other. To partake in the vilification is to contribute to a world of spectacularized impressions, which not only control one's attention but also discredit the value of an alternative future; to keep operating under the current values of civility and security on one side, and barbarity and threat on the Other side, is to remain living in a present-continuous "*truth* of modernity."

The inability to grasp the immensity of the violence being implemented circumscribes protests since the underlying agenda of exercising humanism, that is, violence, is in the name of border security and stable economies. In this whole process, technology is perceived as a neutral apparatus as if utilized without the involvement of a human agent; despite that same technology – the mass media, for example – being drafted to frame and evaluate violence and to directly influence our perception of the Other. The nature of invisible violence is highlighted in George Orwell's *Animal Farm*, where freedom is contingent on muting thoughts and ideas that challenge the oppressive political structures.[10] This stifling, according to Orwell, is undertaken voluntarily. That such a choice can be made willingly indicates that freedom comes at a price; individual thoughts and desires are possible only if consistent with the overarching governing structures. Correspondingly, to aspire to national security and freedom requires partnering with the dominant global powers, such as NATO, despite its despotic impositions.

The current global culture of security protectionism and borders enforced in the present-continuous is bound to shift and change; all systems and networks come to an end or are recalibrated. Whether the end is imminent or distant will depend on how balkanization is implemented: as a system of parcelization for purposes of creating homogeneous enclaves and enforcing power structures or to bring together different entities for purposes of alternative globalization, heterogeneity and socioeconomic solidarity (as was seen in Yugoslavia). If in pursuit of homogeneity, this would lead to economic debt and of degradation of the Other in the name of 'normality' – it is a position of freedom *from* difference. This type of freedom has been administered since the formation of ethnically homogeneous nation states. In the alternative scenario, it could lead to destructuring the current systems and the potential for a new culture of freedom *of* difference.

Other spaces, other times

The fortification of camp Europe and Anglo-America begs the question of how to transform thinking and policies in response to the current migrant and refugee crisis. In other words, how do we engage with the question of security, but not at the price of vilifying the Other through association with a threat and an economic burden or of making security contingent on immobilizing any thought beyond those deemed worthy by MIME?

Considering that the European border against refugees and the displaced is drawn along the Mediterranean coastline and the edge of the Western Balkans, the pressing question is how Balkanism, and balkanization as it was implemented during the 20th Century Yugoslavia, may be used to rethink borders and ownership. The outside does exist, though not from the perspective of an absolute and all-knowing understanding. To stand outside is simply to ask different kinds of questions and derive different kinds of answers. In other words, to formulate questions in a more open, yet precise, way so that the responses generate something that previously did not exist. These questions and answers must go beyond the balkanization associated with the Enlightenment: from the formation of ethnically homogeneous nation states to scientific classification, to the separation of people on the basis of whether they share 'civil' values, to identifying those who are delinquents and therefore require vilification and punishment. To step outside is to perceive beyond the construct of fear of an Other which undermines social, economic and political liberties. Moreover, to remove fear is to remove the preconceptions of Other as the violator, including the preconception that NATO's interventions, such as the 1999 Operation Allied Force, are permissible for purposes of protecting the values of the Euro-Atlantic community.[11] Hence, the question is how to open up our thinking and pose critically precise questions in a society of image and information simulation; when freedom is better understood if posed in terms of how to keep control at a tolerable level.

Thinking alternatives needs to be reproblematized because certain aspects also need to be clarified in terms of what is to be seen as an alternative. Often what is presented as an alternative is nothing more than a type of disciplinary practice disguised through the invocation of topical catch-phrases and ready-made solutions enacted and invoked in the name of terrorism. Yet terrorism accommodates racism and militarism; it makes racism and militarism permissible, as it is invoked for purposes of security, defence and protection. To place terrorism, racism and militarism side by side does not mean that they are the same but that there is an implicit relationship between them, the common denominator being exploitation of the Other and an opposition to inclusive heterogeneity.

The alternative needs instead to have a new rationale concerning security, an innovative economic and social logic and a different organization of time in terms of work and leisure. To do so, there is a need to think alternatives in a way that is removed from the current rhetoric imposed by MIME. The alternative also needs to bracket out the rhetoric of fear and danger; emotions that bookend every news footage of threat and the need for border security. Neither is the alternative the conception of utopia as a non-place or as an unattainable space that exists only in theory and as an abstraction; that which will never be reached in practice. Instead, it is about thinking and implementing ways that facilitate opportunities for even access to resources and spaces for all. This is not to say that we are all equal, since we are all different, but that equality should be measured in terms of equity and affordance of accessibility. Hence, if we have a sense of responsibility to the medium- and long-term future, then the fears that have been imposed need to be eliminated. For historiography to be activated, it needs to start operating in a way that allows the human to remain undefined, open and certainly not to follow the herd of cult figures, particularly in the discourse of politics and international human rights, since the human – according to Judith Butler – tends to be "defined in advance, in terms that are distinctively western, very often American, and, therefore, partial and parochial."[12]

One of the lessons from the Yugoslav socialist model – at least in terms of economics – is that any new culture hoping to establish an alternative model to the market economy cannot be created in dependence on that very system. This was clearly demonstrated during the remaking of Yugoslavia post-WWII; its success – its eventual failure – was founded on foreign financial aid as well as quick returns and high interest rates. In other words, for a more socioeconomically and politically aware culture to arise, debt in the way it is applied today needs to be abolished or at the very least revaluated.

Another lesson from Titoism, highlighted in its participatory role in the Non-Aligned Movement, is that the alternative still needs to work with the existing structure. The alternative political system still maintained dialogue with the USSR and the West (though in varying degrees during its existence), whilst forging an alternative third path of non-alignment. The catch-22, and one that certainly brought the alternative path to a halt due to various economic loans including later international involvement to do with the dissolution of the SFRY, is the possibility of working with existing systems without being swallowed by them. This is not to say that the period of the SFRY should be perceived as the golden model or the only tool kit of lessons since the events from the past are certainly different from those of the present. It is more that thinking balkanization as deployed in Yugoslavia is to consider more open models and emergent policies to do with borders whereby territories are to be defined in terms of constituent nations and national minorities for purposes of inclusivity and equitable access to systems and resources.

From an urban perspective and by using the case study of New Belgrade – whose urban planning is not only physically incomplete to this day, but also where the variety and multiplicity of urban plans produced up until the 1950s for New Belgrade's large-scale construction reveal that notions of a particular ideal were not set – may proffer the opportunity to suggest that being democratic is possible only if democracy is understood as being a project in process. If it is a process, then democracy is implicitly congruent with change and evolution, treating neither history nor memory as eternal but as fragments from which arise both decay and birth. Effectively, a country or indeed any process of participation and decision-making cannot be called democratic if its values and history are not open for contestation; the ascription of democracy – rather than democracy-in-process – signifies that elements of totalitarianism are present. Thinking crisis is thinking collaboration via multiplicity, and where alternatives certainly need to differ from current practices that invoke dehumanization in the name of rights, where violence is enacted in the name of security, and where people are denied access in the name of territory and lack of historical belonging.

Addressing alternatives on a global level through the type of balkanization implemented in Yugoslavia is to stand outside this history; where being human is refusing this kind of humanity. And, most important, to think crisis is not to think emergency or the need to make rapid and often hasty decisions. Instead, thinking crisis is an opportunity to alter the way in which decisions have been made thus far. Bauman rightly points to the etymological relation of crisis to the term "criterion" which is implicated in making a decision rather than the identification of a "catastrophe" or "disaster" in which the term is situated today.[13] This decision needs to propose an alternative to the current political landscape of balkanization – exemplified by Brexit – where distinct territories are reformed for purposes of accommodating only singular values. Neither is it about forming unions or organizations – such as the EU or NATO – where participation and membership accedes to imposed regulations towards the Other. Instead, by using Balkanism as a descriptive and evaluative term to do with liminality, flux and frictional multiplicity in which all Western values and conventions are placed into question, spatializing balkanization is more about thinking and implementing programmes and policies where rhizomatic and, of other spaces and other times; where values and conventions are multiple and can be challenged.

Notes

1 Refer to Mike Davis, "Fortress Los Angeles: The Militarization of Urban Space," in *Variations on a Theme Park: The New American City and End of Public Space*, ed. Michael Sorkin (New York: Hill and Wang, 1992), 154–180; Michael Sorkin, "Urban Warfare: A Tour of the Battlefield," in *Cities, War and Terrorism: Towards and Urban Geopolitics*, ed. Stephen Graham (Oxford: Blackwell, 2004), 251–263; Stephen Graham, *Cities Under Siege: The New Military Urbanism* (London: Verso, 2011); Loïc Wacquant, "The Militarization of Urban Marginality: Lessons from the Brazilian Metropolis," *International Political Sociology* 2:1 (2008): 56–74; and Marcelo Lopes de Souza, "The Brave New (Urban) World of Fear and (Real or Presumed) Wars," *City* 14:4 (2010): 457–463.

2 Giorgio Agamben, *Homo Sacer: Sovereign Power and Bare Life* (Stanford: Stanford University Press, 1998).

3 Paul Latawski and Martin A. Smith, eds., *The Kosovo Crisis and the Evolution of Post-Cold War European Security* (Manchester: Manchester University Press, 2003), 66.

4 VU University Amsterdam, *Human Costs of Border Control*, <www.borderdeaths.org/>.

5 European Council. Council of the European Union, "Address by President Donald Tusk to the European Parliament on the Informal Meeting of Heads of State or Government of 23 September 2015," 6 October 2015, <www.consilium.europa.eu/en/press/press-releases/2015/10/06/tusk-address-european-parliament-informal-euco-september/>.

6 Martin Heidegger, *Being and Time* (Hoboken: Wiley-Blackwell, 1978).

7 Michael Hardt, "The Withering of Civil Society," in *Deleuze & Guattari: New Mappings in Politics, Philosophy, and Culture*, eds. Eleanor Kaufman and Kevin Jon Heller (Minneapolis: University of Minnesota Press, 1998), 28.

8 Zygmunt Bauman, *Modernity and the Holocaust* (Ithaca: Cornell University Press, 1989), 6.

9 Kriss Ravetto, *The Unmaking of Fascist Aesthetics* (Minneapolis: University of Minnesota Press, 2001), 8.

10 George Orwell, *Animal Farm* (London: Penguin, 2000).

11 Latawski and Smith, *The Kosovo Crisis and the Evolution of Post-Cold War European Security*, 166.

12 Judith Butler, *Undoing Gender* (New York: Routledge, 2004), 36–37. Despite the title suggesting that Butler is addressing notions of gender, the underlying narrative generally deals with the condition and the constituencies of being human. Also see Ben Golder, *Foucault and the Politics of Rights* (Palo Alto: Stanford University Press, 2015).
13 Zygmunt Bauman, *In Search of Politics* (Cambridge: Polity, 1999), 140.

Bibliography

Abramowitz, Morton. "Preface, The Other Balkan Wars: A 1913. Carnegie Endowment Inquiry," in *Retrospect with a New Introduction and Reflections of the Present Conflict*, edited by George Kennan. Washington, DC: Carnegie Endowment for International Peace, 1993.

Abujidi, Nurhan and Han Verschure. "Military Occupation as Urbicide by 'Construction and Destruction': The Case of Nablus, Palestine." *The Arab World Geographer/Le Géographe du monde arabe* 9:2 (2006): 126–154.

Adams, Nicholas. "Architecture as the Target." *Journal of the Society of Architectural Historians* 52 (1993): 389–390.

Adler, Jonathan H. "Climate Balkanization: Dormant Commerce and the Limits of State Energy Policy." *LSU Journal of Energy Law and Resources* 3:1 (Fall 2014): 153–194.

Agamben, Giorgio. *State of Exception*. Chicago: The University of Chicago Press, 2005.

Agamben, Giorgio. "Security and Terror." *Theory and Event* 5:4 (2002): 1–2.

Agamben, Giorgio. *Homo Sacer: Sovereign Power and Bare Life*. Stanford: Stanford University Press, 1998.

Agerman, Lorenc, Eduard Frojdam and Djan Gildji. *Beograd Gazela: Vodic kroz Sirotinjsko Naselje/Belgrade Gazela: A Guide Through the Poverty Settlement*. Beograd: Rende, 2009.

Aguilera, Antonio. "Ethics and Violence." In *Globalization, Political Violence and Translation*, edited by Esperanza Bielsa and Christopher Hughes, 72–88. Basingstoke: Palgrave Macmillan, 2009.

Aldrich, Howard E. and Roger Waldinger. "Ethnicity and Entrepreneurship." *Annual Review of Sociology* 16 (August 1990): 111–135.

Alesina, Alberto and Eliana La Ferrara. "Participation in Heterogeneous Communities." *Quarterly Journal of Economics* 115:3 (2000): 847–904.

Alfirević, Đorđe and Sanja Simonović Alfirević. "Urban Housing Experiments in Yugoslavia 1948–1970." *Spatium* 34 (December 2015): 1.

Allen, Bob. "Why Kosovo? The Anatomy of a Needless War." *Canadian Centre for Policy Alternatives – BC Office* (July 1999): 11.

Almond, Michael. "Belgrade Open for Business as Usual." *The Guardian*. 16 May 1999.

Alneng, Victor. "The Modern Does Not Cater for Natives: Travel Ethnography and the Conventions of Form." *Tourist Studies* 2:2 (2002): 119–142.

Alneng, Victor. " 'What the Fuck Is a Vietnam?': Touristic Phantasms and the Popcolonization of (the) Vietnam (War)." *Critique of Anthropology* 22:4 (2002): 461–489.

Amnesty International. "Roma Families in Belgrade Face Forced Eviction by Government Agency." 2 November 2011. <www.amnesty.org/en/news/roma-families-belgradeface-forced-eviction-government-agency-2011-11-02>.

Amnesty International. "Federal Republic of Yugoslavia (FRY)/NATO: 'Collateral Damage' or Unlawful Killings? Violations of the Laws of War by NATO During Operation Allied Force." 5 June 2000. <www.amnesty.org/en/documents/eur70/018/2000/en/>.

Anderson, Peter. "Air Strike: NATO Astride Kosovo." In *The Kosovo Crisis: The Last American War in Europe?* edited by Tony Weymouth and Stanley Henig. London: Reuters, 2001.

Arkin, William M. "Smart Bombs, Dumb Targeting?" *The Bulletin of the Atomic Scientists* 56:3 (2000): 46–53.

Baillie, Britt. "Capturing Facades in 'Conflict-Time': Structural Violence and the (Re)construction Vukovar's Churches." *Space and Polity* 17:3 (2013): 300–319.

Baillie, Britt. "Memorialising the 'Martyred City': Negotiating Vukovar's Wartime Past." In *Locating Urban Conflicts*, edited by Wendy Pullen and Britt Baillie, 115–131. London: Palgrave Macmillan, 2013.

Baillie, Britt. "Vukovar's Divided Memory: The Reification of Ethnicity Through Memorialisation." *Divided Cities/Contested States Working Paper Series/Working Paper 25 – Conflict in Cities and the Contested State*. Department of Architecture, University of Cambridge, (2012): 1–22. <www.conflictincities.org/PDFs/WorkingPaper25(Divided Memory).pdf>.

Baker, Catherine. "Music as a Weapon of Ethnopolitical Violence and Conflict: Processes of Ethnic Separation During and After the Break-Up of Yugoslavia." *Patterns of Prejudice* 47:4–5 (2013): 409–429.

Bakic-Hayden, Milica. "Nesting Orientalisms: The Case of the Former Yugoslavia." *Slavic Review* 54:4 (1995): 917–931.

Bakic-Hayden, Milica and Robert Hayden. "Orientalist Variations on the Theme 'Balkans': Symbolic Geography in Recent Yugoslav Cultural Politics." *Slavic Review* 51:1 (1992): 1–15.

Ballantyne, Andrew. "Architecture as Evidence." In *Rethinking Architectural Historiography*, edited by Dana Arnold and Belgin Turan Ozkaya, 36–50. London: Routledge, 2006.

Baltimore, Frederick Calvert Baron. *A Tour to the East, in the Years 1763 and 1764. With Remarks on the City of Constantinople and the Turks*. London: W. Richardson and S. Clark, 1767.

Barakat, Sultan. "City War Zones." *Urban Age* 5:4 (Spring 1998): 10–16.

Barlovac, Bojana. "Reconstruction of Main Belgrade Bridge Kicks Off." *BalkanInsight.com*. 1 July 2010. <www.balkaninsight.com/en/ main/news/29160/>.

Barnett, Michael. *Empire of Humanity*. New York: Cornell University Press, 2013.

Baudrillard, Jean. *The Gulf War Did Not Take Place*. Sydney: Power Publications, 1995.

Bauman, Zygmunt. *Liquid Times*. Cambridge: Polity Press, 2007.

Bauman, Zygmunt. *Liquid Fear*. Cambridge: Polity Press, 2006.

Bauman, Zygmunt. "Living (Occasionally Dying) Together in an Urban World." In *Cities, War and Terrorism: Towards an Urban Geopolitics*, edited by Stephen Graham, 110–120. Oxford: Blackwell Publishing, 2004.

Bauman, Zygmunt. *Liquid Modernity*. Cambridge: Polity Press, 2000.

Bauman, Zygmunt. *In Search of Politics*. Cambridge: Polity, 1999.

Bauman, Zygmunt. *Modernity and the Holocaust*. Ithaca: Cornell University Press, 1989.

BBC News. "Nato Missile Video 'No Distortion.'" Friday, 7 January 2000. <http://news.bbc.co.uk/1/hi/world/europe/594800.stm>.

Benjamin, Walter. "Theses on the Philosophy of History." In *Illuminations*, edited and with an introduction by Hannah Arendt. New York: Shocken, 1977.

Bennett, Christopher. *Bosnia's Paralysed Peace*. London: Hurst, 2016.

Benson, Leslie. *Yugoslavia: A Concise History*. Basingstoke: Palgrave Macmillan, 2004.

Ben-Ze'ev, Efrat and Ruth Ginio, eds. *Shadows of War: The Social History of Silence in the Twentieth Century*. Cambridge: Cambridge University Press, 2010.

Bergson, Henri. *Matter and Memory*. New York: Zone Books, 2005.

Berman, Marshall. *Adventures in Marxism*. New York: Verso, 1999.

Berman, Marshall. "Falling Towers: City Life After Urbicide." In *Geography and Identity: Living and Exploring Geopolitics of Identity*, edited by Dennis Crow, 172–192. Washington, DC: Maisonneuve Press, 1996.

Berman, Marshall. "Among the Ruins." *New Internationalist* 178. December 1987. <http://newint.org/features/1987/12/05/among>.

Bevan, Robert. *The Destruction of Memory – Architecture at War*. London: Reaktion Books, 2007.

Bhagwati, Jagdish. *In Defense of Globalization*. Oxford: Oxford University Press, 2007.

Bielsa, Esperanza and Christopher W. Hughes, eds. *Globalization, Political Violence and Translation*. Basingstoke: Palgrave Macmillan, 2009.

Bjelić, Dušan I. "Introduction: Blowing Up the 'Bridge.'" In *Balkan as Metaphor*, edited by Dušan I. Bjelić and Obrad Savić, 1–23. Cambridge, MA: MIT Press, 2002.

Bjelić, Dušan I. and Obrad Savić. *Balkan as Metaphor*. Cambridge, MA: MIT Press, 2002.

Bjelsa, Esperanca. "Globalization, Political Violence and Translation: an Introduction." In *Globalization, Political Violence and Translation*, 1–25. Basingstoke: Palgrave Macmillan, 2009.

Blagojević, Ljiljana. "Urban Regularisation of Belgrade, 1867: Trace vs. Erasure." *SAJ* 1 (2009): 27–44.

Blagojević, Ljiljana. *Novi Beograd: Osporeni Modernizam*. Beograd: Zavod za udžbenike, Zavod zaštitu spomenika kulture grada, Arhitektonski fakultet, 2007.

Blagojević, Ljiljana. "New Belgrade: The Capital of No-City's Land." *Art-e-Fact, Strategies of Resistance* 4 (December 2005). <http://artefact. mi2.hr/_a04/lang_en/theory_blagojevic_en.htm>.

Bobic, Nikolina. "Belgrade in Formation(s): Dobrovic's Generalstab Complex." *Fabrications* 21:1 (2012): 7–28.

Bogdanovic, Bogdan. "The City and Death." In *Balkan Blues: Writing Out of Yugoslavia*, edited Joanna Labon, 37–74. Evanston: Northwestern University Press, 1995.

Bougarel, Xavier. "Yugoslav Wars: The 'Revenge of the Countryside' Between Sociological Reality and Nationalist Myth." *East European Quarterly* 33:2 (1999): 157–175.

Bovard, James. "Kosovo Déjà vu." *The Future of Freedom Foundation*. 1 July 2003. <http://fff.org/explore-freedom/article/kosovo-dj-vu/>.

Brenner, Neil and Roger Keil, eds. *The Global Cities Reader*. New York: Routledge, 2005.

Buchanan, Ian. "Treatise on Militarism." In *Deleuze and the Contemporary World*, edited by Ian Buchanan and Adrian Parr, 21–42. Edinburg: Edinburg University Press, 2006.

Bunjac, Vladimir. *Jedan Vek Beogradskog Vodovoda i Kanalizacije 1892–1992*. Beograd: BMS, 1992.

Butler, Judith. *Frames of War: When Is Life Grievable?* New York: Verso, 2016.

Butler, Judith. *Undoing Gender*. New York: Routledge, 2004.

Butler, Octavia E. *Parable of the Sower*. New York: Four Wall Eight Windows, 1993.

Calame, Jon and Esther Charlesworth. *Divided Cities*. Philadelphia, PA: University of Pennsylvania Press, 2009.

Campbell, David et al. "Urbicide: The Killing of Cities." *Theory & Event* 10:2 (2007). <https://muse.jhu.edu/login?auth=0&type=summary&url=/journals/theory_and_event/v010/10.2campbell.html>.

Castells, Manuel. *The Rise of the Network Society*. Oxford: Wiley-Blackwell, 2000.

Chomsky, Noam. *Failed States*. New York: Metropolitan Books, 2006.

Chomsky, Noam. *The New Military Humanism*. London: Pluto Press, 1999.

Christie, Agatha. *The Secret of Chimneys*. London: The Bodley Head, 1925.

Civallero, Edgardo. "When Memory Turns into Ashes . . . Memoricide During the XX Century." *Information for Social Change* (Summer 2007): 1–13.

Clinton, Bill. "Clinton Rodney King Trial Remarks." *C-SPAN.org*. April 17, 1993 (user-created clip, July 19, 1993). <www.c-span.org/video/?c4459701/clinton-rodney-king-trial-remarks>.

Clinton, William J. "Remarks at the Legislative Convention of the American Federation of State, County, and Municipal Employees." *The American Presidency Project*. 23 March 1999. <www.presidency.ucsb.edu/ws/?pid=57294>.

Coker, Christopher. *Humane Warfare*. London: Routledge, 2001.

Council of Europe Committee of Ministers. "Recommendation Rec (2001) 15 of the Committee of Ministers to Member States on History Teaching in Twenty-First-Century Europe (Adopted by the Committee of Ministers on 31 October 2001). <https://pestalozziprogrammelibrary.wordpress.com/2015/01/04/recommendation-rec200115-of-the-committee-of-ministers-to-member-states-on-history-teaching-in-twenty-first-century-europe/>.

Coward, Martin. *Urbicide: The Politics of Urban Destruction*. New York: Routledge, 2009.

Coward, Martin. " 'Urbicide' Reconsidered." *Theory & Event* 10:2 (2007). <http://muse.jhu.edu.ezproxy1.library.usyd.edu.au/journals/theory_and_event/v010/10.2coward.html>.

Coward, Martin. "Against Anthropocentrism: The Destruction of the Built Environment as a Distinct Form of Political Violence." *Review of International Studies* 31:3 (2006): 419–437.

Coward, Martin. "Urbicide in Bosnia." In *Cities, War and Terrorism*, edited by Stephen Graham, 154–172. Oxford: Blackwell, 2004.

Čvoro, Uroš. *Turbo Folk and National Identity in Former Yugoslavia*. London: Ashgate, 2014.

Davis, Mike. *City of Quartz*. London: Verso, 2006.

Davis, Mike. "Fortress Los Angeles: The Militarization of Urban Space." In *Variations on a Theme Park: The New American City and End of Public Space*, edited by Michael Sorkin, 154–180. New York: Hill and Wang, 1992.

Dear, Michael. "Los Angeles and the Chicago School: Invitation to a Debate." *City & Community* 1:1 (2002): 5–32.

Debeljak, Aleš. "On the Balkans." *Balkans Project*.

Dedijer, Vladimir. *The Yugoslav Auschwitz and the Vatican: The Croatian Massacre of the Serbs During World War II*. London: Prometheus Books, 1992.

Deleuze, Gilles. *Negotiations, 1972–1990*. New York: Columbia University Press, 1995.

Deleuze, Gilles. *The Logic of Sense*. New York: Columbia University Press, 1990.

Deleuze, Gilles and Felix Guattari. *Anti-Oedipus: Capitalism and Schizophrenia*. New York: Penguin Books, 2009.

Deleuze, Gilles and Felix Guattari. *A Thousand Plateaus: Capitalism and Schizophrenia*. Minneapolis, MN: Minnesota University Press, 1988.

Der Derian, James. *Virtuous War: Mapping the Military-Industrial-Media-Entertainment Network*. Routledge: London, 2009.

Der Derian, James. "Global Swarming, Virtual Security and Bosnia." *The Washington Quarterly* 19:3 (1996): 45–56.

Der Derian, James. "International Theory, Balkanisation and the New World Order." *Millennium Journal for International Studies* 20:3 (1991): 485–506.

Derrida, Jacques. "Des Tours de Babel." In *Difference in Translation*, edited by Joseph E. Graham, 165–207. Ithaca: Duke University Press, 1985.

de Souza, Marcelo Lopes. "The Brave New (Urban) World of Fear and (Real or Presumed) Wars." *City* 14:4 (2010): 457–463.

de Tocqueville, Alexis. *Democracy in America*. Vol. 1. London; Vintage, 1990.

de Tocqueville, Alexis. *Democracy in America*. Vol 2. London: Vintage, 1990.

Dinkić, Mlađan, ed. *Final Account Economic Consequences of NATO Bombing: Estimate of the Damage and Finances Required for the Reconstruction of Yugoslavia*. Beograd: Stubovi Kulture, 1999.

Dizdarević, Zlatko. *Sarajevo: A War Journal*. New York: Fromm International, 1997.

Dobrović, Nikola. *Savremena arhitektura 3*. Beograd: Zavod za izdavanje udžbenika Socijaličke Republike Srbije, 1963.

Doel, Marcus A. *Geographies of Violence*. New York: Sage, 2017.

Du Bois, W. E. B. *The World and Africa: Colour and Democracy, Colour and Race*. New York: Oxford University Press, 2007.

Dudley, Michael Q. "Revisiting Cold War Ideology in the Secure City: Towards a Political Economy of Urbicide." *Theory & Event* 10:2 (2007).

Economist. "Gladstone's Ghost." 27 May 1999. <www.economist.com/britain/1999/05/27/gladstones-ghost>.

Economist Intelligence Unit. "Country Report: Yugoslavia (Serbia-Montenegro)." 2000.

Energy Community Homepage. *Chapter 15: Energy – European Commission Opinion on Serbia's Application for Membership of the European Union 12 Oct 2011*.

European Council. Council of the European Union. "Address by President Donald Tusk to the European Parliament on the Informal Meeting of Heads of State or Government of 23 September 2015." 6 October 2015. <www.consilium.europa.eu/en/press/press-releases/2015/10/06/tusk-address-european-parliament-informal-euco-september/>.

European Union External Action. "Security Union: A European Travel Information and Authorisation System – Questions & Answers." 15:11 (17 November 2016). <https://eeas.europa.eu/headquarters/headquarters-homepage/15205/security-union-european-travel-information-and-authorisation-system-questions-answers_en>.

Faber, Mient Jan. "Cold Wars and Frozen Conflicts: The European Experience." In *Global Insecurity: Restructuring the Global Military Sector v.3*, edited by Mary Kaldor, 53–95. London: Pinter, 2000.

Fassin, Didier. *Humanitarian Reason: A Moral History of the Present Times*. Berkeley: University of California Press, 2011.

Fassin, Didier. "Another Politics of Life Is Possible." *Theory, Culture & Society* 26:5 (2009): 44–60.

Fleming, Katherine E. "*Orientalism*, the Balkans, and Balkan Historiography." *The American Historical Review* 105:4 (2000): 1218–1233.

Forde, Susan. "The Bridge on the Neretva: Stari Most as a Stage of Memory in Post-Conflict Mostar, Bosnia – Herzegovina." *Cooperation and Conflict* 51:4 (2016): 467–483.

Forensic Architecture and Grupa Spomenik. "Living Death Camps." In *Forensis: The Architecture of Public Truth, a Project by Forensic Architecture, Centre for Research Architecture, Department of Visual Cultures*. Goldsmiths, University of London. Berlin: Sternberg Press, 2014.

Forty, Adrian, ed. *The Art of Forgetting*. Oxford: Berg Publishers, 2001.

Foucault, Michel. "Of Other Spaces (1967)." In *Heterotopia and the City: Public Space in a Postcivil Society*, edited by Michiel Dehanene and Lieven De Cauter, 13–29. London: Routledge, 2008.

Foucault, Michel. *Security, Territory, Population: Lectures at the Collège de France 1977–1978*, edited by Michel Senellart. New York: Palgrave Macmillan, 2007.

Foucault, Michel. *The Archaeology of Knowledge and the Discourse on Language*. New York: Vintage, 1982.

Foucault, Michel. *Discipline and Punish: The Birth of the Prison*. New York: Pantheon Books, 1977.

Fouskas, Vassilis and Bulent Gökay. *The New American Imperialism*. Westport: Greenwood Political Publishing, 2005.

Fox, Gregory H. and Brad R. Roth, eds. *Democratic Governance and International Law*. Cambridge: Cambridge University Press, 2000.

Fregonese, Sara. "The Urbicide of Beirut? Geopolitics and the Built Environment in the Early Phases of the Lebanese Civil War (1975–76)." *Political Geography* 28:5 (2009): 309–318.

Friedman, Thomas L. "Foreign Affairs; Stop the Music." *New York Times*. 23 April 1999. <www.nytimes.com/1999/04/23/opinion/foreign-affairs-stop-the-music.html>.

Gagnon Jr, V. P. (Chip). "Yugoslavia in 1989 and After." *Nationalities Papers* 38:1 (2010): 23–39.

Gajević, Ljubica. "New Belgrade Urban Fortunes: Ideology and Practice Under the Patronage of State and Market." Master's Dissertation, University of Catalonia, Barcelona, 2011.

Gallis, Paul. "NATO's Decision-Making Procedure." *CRS Report for Congress*. Updated 5 May 2003. <www.au.af.mil/au/awc/awcgate/crs/rs21510.pdf>.

Gati, Charles. "From Sarajevo to Sarajevo." *Foreign Affairs* 71:4 (Autumn 1992).

Gibbs, David N. *First Do No Harm: Humanitarian Intervention and the Destruction of Yugoslavia*. Nashville: Vanderbilt University Press, 2009.

Gladstone, Rick. "Roma Sickened in U.N. Camps Are Still Waiting for Redress." *New York Times*. 18 April 2017.

Glennon, Michael J. *Limits of Law, Prerogatives of Power: Interventionism After Kosovo*. London: Palgrave Macmillan, 2001.

Glenny, Misha. "Only in the Balkans." *London Review of Books* 21:9 (1999): 12–14.

Glenny, Misha. *The Fall of Yugoslavia*. London: Penguin, 1996.

Gligorijevic, Zaklina. "Can New Developments and City Identity Grow in Harmony? – The Quest for a Successful Public Space Design for New Belgrade." In *Proceedings of 42nd ISoCaRP Congress 2006, Cities Between Integration and Disintegration, September 14–17, 2006*. Turkey: Yıldız Technical University, Faculty of Architecture, Department of City and Regional Planning, Barbaros Bulvari, Besiktas. <www.isocarp.net/Data/case_studies/831.pdf>.

Gocić, Goran. "Symbolic Warfare: Nato Versus the Serbian Media." In *Degraded Capability: The Media and the Kosovo Crisis*, edited by Philip Hammond and Edward S. Herman, 88–97, 2000. London: Pluto Press.

Golder, Ben. *Foucault and the Politics of Rights*. Palo Alto: Stanford University Press, 2015.

Goldsworthy, Vesna. *Inventing Ruritania: The Imperialism of the Imagination*. London: Hurst, 2013.

Goldsworthy, Vesna. Interview – 'Not Quite European . . . Scrutinizing the Balkan Neo-Colonialism.' *Eurozine*. 9 July 2004. <www.eurozine.com/not-quite-european-scrutinizing-the-balkans-neo-colonialism/>.

Goonewardena, Kanishka. "Marxism and Everyday Life: On Henri Lefebvre, Guy Debord and Some Others." In *Space, Difference, Everyday Life: Reading Henri Lefebvre*, edited by Kanishka Goonewardena et al., 117–133. London: Routledge, 2008.

Goudkamp, James. *Tort Law Defences*. Oxford: Hart Publishing, 2013.

Gourgouris, Stathis. *Dream Nation: Enlightenment, Colonization, and the Institution of Modern Greece*. Palo Alto: Stanford University Press, 1996.

Graham, Stephen. *Cities Under Siege: The New Military Urbanism*. London: Verso, 2011.

Graham, Stephen. "Disruption by Design: Urban Infrastructure and Political Violence." In *Disrupted Cities: When Infrastructure Fails*, edited by Stephen Graham, 111–131. New York: Routledge, 2010.

Graham, Stephen. "Demodernizing by Design: Everyday Infrastructure and Political Violence." In *Violent Geographies: Fear, Terror, and Political Violence*, edited by Derek Gregory and Allan Pred, 309–329. New York: Routledge, 2006.

Graham, Stephen. "Constructing Urbicide by Bulldozer in the Occupied Territories." In *Cities, War and Terrorism: Towards and Urban Geopolitics*, edited by Stephen Graham, 192–214. Oxford: Blackwell Publishing, 2004.

Graham, Stephen. "Postmortem City: Towards an Urban Geopolitics." *City* 8:2 (2004): 165–196.

Graham, Stephen. "Lessons in Urbicide." *New Left Review* 19 (January–February 2003). <https://newleftreview.org/II/19/stephen-graham-lessons-in-urbicide>.

Graham, Stephen and Lucy Hewitt. "Getting off the Ground: On the Politics of Urban Verticality." *Progress in Human Geography* 37:1 (2012): 72–92.

Greenwall, H. T. "Head of Great Industry Predicts the 'Balkanization of Europe.'" *The New York Times*. 20 December 1918.

Grubačić, Andrej. *Don't Mourn, Balkanize!* Oakland: PM Press, 2010.

Hammond, Andrew. "Balkanism in Political Context: From the Ottoman Empire to the EU." *Westminster Papers in Communication and Culture* 3:3 (2006): 6–26.

Hammond, Andrew, ed. *The Balkans and the West: Constructing the European Other, 1945–2003*. Farnham: Ashgate, 2004.

Hammond, Philip. *Framing Post-Cold War Conflicts*. Manchester: Manchester University Press, 2011.

Hanafi, Sara. "Explaining Spacio-Cide in the Palestinian Territory: Colonization, Separation, and State of Exception." *Current Sociology* 61:2 (2012): 190–205.

Hardt, Michael. "The Withering of Civil Society." In *Deleuze & Guattari: New Mappings in Politics, Philosophy, and Culture*, edited by Eleanor Kaufman and Kevin Jon Heller, 23–40. Minneapolis: University of Minnesota Press, 1998.

Hardt, Michael and Antonio Negri. *Multitude: War and Democracy in the Age of Empire*. London: Penguin Books, 2004.

Hareven, Tamara K. *Amoskeag: Life and Work in an American Factory-City*. Hanover, NH: University Press of New England, 1978.

Harman, Edward S. "The Political History of Sham Justice. Carla Del Ponte Addresses Goldman Sachs on Justice and Profits." *GlobalResearch*. 20 November 2005. <www.globalresearch.ca/the-hague-tribunal-the-political-economy-of-sham-justice/1294>.

Hartmann, Kristen M. "Fragmentation and Forgetting: Sarajevo's Vijećnica." *International Journal of Heritage Studies* 22:4 (2016): 312–324.

Harvey, David. *Justice, Nature and the Geography of Difference*. Hoboken: Wiley-Blackwell, 1996.

Hedges, Chris. "Kosovo's Next Masters?" *Foreign Affairs* 78:3 (May/June 1999). <www.foreignaffairs.com/articles/europe/1999-05-01/kosovos-next-masters>.

Heidegger, Martin. "Building Dwelling Thinking." In *Basic Writings*. London: Routledge, 2010.

Heidegger, Martin. *Being and Time*. Hoboken: Wiley-Blackwell, 1978.

Helms, Elisaa. "East and West Kiss: Gender, Orientalism, and Balkanism in Muslim-Majority Bosnia-Herzegovina." *Slavic Review* 67:1 (Spring 2008): 88–119.

Herscher, Andrew. *Violence Taking Place: The Architecture of the Kosovo Conflict*. Stanford: Stanford University Press, 2010.

Herscher, Andrew. "Warchitectural Theory." *Journal of Architectural Education* 61:3 (2008): 35–43.

Herscher, Andrew. "Urbicide, Urbanism and Urban Destruction in Kosovo." *Theory & Event* 10:2 (2007).

Hewitt, Kenneth. "Proving Grounds of Urbicide: Civil and Urban Perspectives on the Bombing of Capital Cities." *ACME: An International e-Journal for Critical Geographies* 8:2 (2009): 340–375.

Heylin, Michael. "NATO Bombs Take Out Chemical Complex." *Chemical and Engineering News* 77:19 (1999).

Hill, Jonah. "Internet Fragmentation: Highlighting the Major Technical, Governance and Diplomatic Challenges for U.S. Policy Makers." Report. Belfer Center for Science and International Affairs (BCSIA) – Harvard Kennedy School (HKS). 2012.

Hirst, Paul. *Space and Power*. Cambridge: Polity Press, 2005.

Hobsbawm, Eric J. *Nations and Nationalism Since 1780*. Cambridge: Cambridge University Press, 1990.

Hoffman, Frank G. "Complex Irregular Warfare: The Next Revolution in Military Affairs." *Orbis* 50:3 (Summer 2006): 395–411.

Hope, Anthony. *The Prisoner of Zenda*. Bristol: J.W. Arrowsmith, 1894.

Höpken, Wolfgang. "Between Civic Identity and Nationalism: History Textbooks in East Central and Southern Europe." In *Democratic Transition in Croatia: Value Transformation, Education and Media*, edited by Sabrina P. Ramet and Davorka Matic, 163–192. Austin: Texas A&M University Press, 2007.

Huntington, Samuel. "The Clash of Civilizations?" *Foreign Affairs* 72:3 (Summer 1993).

Huxtable, Ada Louise. "Manchester, N.H.: Lessons in Urbicide." In *Goodbye History, Hello Hamburger: An Anthology of Architectural Delights and Disasters*, edited by Ada Louise Huxtable, 102–105. Washington, DC: Preservation Press, 1986.

Huxtable, Ada Louise. *Kicked a Building Lately?* New York: Quadrangle/The New York Times Book Co., 1976.

International Crisis Group. "Milosevic's Aims in War and Diplomacy." *Balkans Report* No. 65. 11 May 1999. <www.crisisgroup.org/europe-central-asia/balkans/serbia/milosevics-aims-war-and-diplomacy>.

James, Patrick. "Tourism and the 'Martyred City': Memorializing War in the Former Yugoslavia." *Journal of Tourism and Cultural Change* 14:3 (2016): 222–239.

Jano, Dorian. "From 'Balkanization' to 'Europeanization': The Stages of Western Balkans Complex Transformations." *L'Europe en formation* 349–350 (2008): 55–69.

Jeffries, Ian. *The Former Yugoslavia at the Turn of the Twenty-First Century: A Guide to the Economies in Transition*. London: Routledge, 2002.

Johnstone, Diana. *Fools' Crusade: Yugoslavia, NATO and Western Delusions*. London: Pluto Press, 2002.

Joksimovich, Vojin. "Militarism and Ecology: NATO Ecocide in Serbia." *Mediterranean Quarterly* 11:4 (2000): 140–160.

Jović, Dejan. *Yugoslavia: A State that Withered Away*. Indiana: Purdue University Press, 2009.

Kaplan, Robert D. *Balkan Ghosts: A Journey Through History*. London: Picador, 2005.

Kardov, Kruno. "Remember Vukovar: Memory, Sense of Place and the National Tradition." In *Battle of Vukovar*, edited by J. Russell and R. Cohn. Edinburg: Bookvika, 2012.

Kaufman, Joyce P. "NATO and the Former Yugoslavia: Crisis, Conflict and the Atlantic Alliance." *The Journal of Conflict Studies* 19:2 (1999). <https://journals.lib.unb.ca/index.php/jcs/article/view/4355/5009>.

Kiossev, Alexander. "The Dark Intimacy: Maps, Identities, Acts of Identification." In *Balkan as Metaphor*, edited by Dušan I. Bjelić and Obrad Savić, 165–191. Cambridge, MA: MIT Press, 2002.

Klein, Naomi. *The Shock Doctrine: The Rise of Disaster Capitalism*. London: Allen Lane, 2007.

Klemenčić, Matjaž and Mitja Žagar. *The Former Yugoslavia's Diverse Peoples: A Reference Sourcebook*. Santa Barbara: ABC-CLIO, 2003.

Kolstø, Pål. " 'Western Balkans' as the New Balkans: Regional Names as Tools for Stigmatisation and Exclusion." *Europe-Asia Studies* 68:7 (2016): 1245–1263.

Komlenović, Uroš. "Belgrade Metro: Lightly Made Promises." *Vreme News Digest Agency*. 5 July 1997. <www.scc.rutgers.edu/serbian_digest/300/t300-6.htm.>

Koolhaas, Rem, Bruce Mau and OMA. *S, M, L, XL*. New York: Monacelli Press, 1995.

Korica, Rajko Lj. and Danilo S. Furundzic. "Metro or Light Rail: Belgrade Transport Proposals." In *Proceedings of Real Corp 2011 Conference on Change for Stability: Lifecycles of Cities and Regions, 18–20 May 2011 Essen*. Essen: Real Corp, 2011. <www.corp.at/archive/CORP2011_173.pdf>.

Kostić, Slobodan. "Simbolika Vecne Vatre." *Vreme 493*. 17 June 2000.

Kostova, Ludmilla. *Tales of the Periphery: The Balkans in nineteenth century British Writing*. Veliko Turnovo: St. Cyril and St. Methodius University Press, 1997.

Koulouri, Christina, ed. *Clio in the Balkans: The Politics of History Education*. Thessaloniki: Center for Democracy and Reconciliation in Southeast Europe, 2002.

Kristeva, Julia. *Powers of Horror: An Essay on Abjection*. New York: Columbia University, 1982.

Krusche, Jürgen. "Bureau Savamala Belgrade." In *Bureau Savamala Belgrade: Urban Research and Practice in a Fast-Changing Neighborhood*, edited by Jürgen Krusche and P. Klaus. Berlin: JOVIS Berlag, 2015.

Kulić, Vladimir. "Prostor u kretanju." *Republika*. 2010. <www.republika.co.rs/476-477/21.html>.

Kulić, Vladimir. "East? West? Or Both? Foreign Perceptions of Architecture in Socialist Yugoslavia." *The Journal of Architecture* 14:1 (2009): 129–147.

Kurth Barth, William. *On Cultural Rights: The Equality of Nations and the Minority Legal Tradition*. Boston: Martinus Nijhoff, 2008.

Kusturica, Emir, dir. *Black Cat, White Cat*. Film. FR Yugoslavia: Komuna, 1998.

Kusturica, Emir, dir. *Underground*. FR Yugoslavia: Komuna, 1995.

Kusturica, Emir, dir. *When Father Was Away on Business*. Yugoslavia: Centar Film/Forum Sarajevo, 1985.

Lahoud, Adrian. "Introduction. Post-Traumatic Urbanism." In *Post-Traumatic Urbanism: Architectural Design* 80:5, edited by Adrian Lahoud et al., 14–23. London: Wiley 2010.

Lalović, Ksenija, Uroš Radosavljević and Zoran Đukanović."Reframing Public Interest in the Implementation of Large Urban Projects in Serbia: The Case of Belgrade Waterfront Project." *Architecture and Civil Engineering* 13:1 (2015): 35–46.

Latawski, Paul and Martin A. Smith. *The Kosovo Crisis and the Evolution of Post-Cold War European Security*. Manchester: Manchester University Press, 2003.

Laursen, Andreas. "NATO, the War over Kosovo, and the ICTY Investigation." *American University International Law Review* 17:4 (2002): 765–814.

Lavrence, Christine. " 'The Serbian Bastille': Memory, Agency, and Monumental Public Space in Belgrade." *Space and Culture* 8:1 (2005): 31–46.

Lazar, Marta Vukotic. *Beogradsko Razdoblje Arhitekte Nikole Dobrovica (1945–1967)*. Beograd: Plato, 2002.

Lazarevic, Eva Vanista, Boris Koruznjak and Mirjana Devetakovic. "Culture Design-Led Regeneration as a Tool Used to Regenerate Deprived Areas. Belgrade – The Savamala Quarter; Reflections on an Unplanned Cultural Zone." *Energy and Buildings* 115 (2016): 3–10.

Leeson, Peter T. "Balkanization and Assimilation: Examining the Effects of State-Created Homogeneity." *Review of Social Economy* LXV:2 (2007): 141–164.

Logan, John R., Richard D. Alba and Thomas L McNulty. "Ethnic Economies in Metropolitan Regions: Miami and Beyond." *Social Forces* 72:3 (1994): 691–724.

Longinovic, Tomislav Z. *Vampire Nation: Violence as Cultural Imaginary*. London: Duke University Press, 2011.

Lukić, Mirjana. "Javni Prostori u Više Nivoa Zeleznicke Stanice 'Vukov Spomenik' – Petnaest godina posle." *Izgradnja* 65:1–2 (2011): 70–81.

Lyotard, Jean Francois. *The Assassination of Experience by Painting, Monory*. London: Black Dog, 1998.

Magaš, Branka. *The Destruction of Yugoslavia*. London: Verso, 1993.

Malcolm, Noel. *Kosovo: A Short History*. New York: New York University Press, 1998.

Mandel, Michael. "Illegal Wars, Collateral Damage, and International Criminal Law." In *Yugoslavia Unraveled: Sovereignty, Self-Determination, Intervention*, edited by Raju G. S. Thomas, 287–317. New York: Lexington Books, 2003.

Manning, Olivia. *Friends and Heroes* (The Balkans Trilogy). Portsmouth, NH: Heinemann, 1965.

Manning, Olivia. *The Spoilt City* (The Balkans Trilogy). Portsmouth, NH: Heinemann, 1962.

Manning, Olivia. *The Great Fortune* (The Balkans Trilogy). Portsmouth, NH: Heinemann, 1960.

Markovich, Slobodan G. *British Perceptions of Serbia and the Balkans*. Paris: Dialogue, 2000.

Martin, Vivian. "Chapter 25: Civilian Casualties of NATO's War on Yugoslavia." *International Action Center*. <https://iacenter.org/warcrime/25_civil.htm>

Massey, Douglas S. and Nancy N. Denton. *America Apartheid*. Cambridge, MA: Harvard University Press, 1993.

Mazower, Mark. *The Balkans*. London: Phoenix Press, 2001.

McNair, Brian. *News and Journalism in the UK*. London: Routledge, 1999.

Meštrović, Stjepan G. *The Balkanization of the West: The Confluence of Postmodernism with Postcommunism*. London: Routledge, 1994.

Miličević, Dušan. "Ruševine koje svima vezu ruke." *Blic*. 30 April 2009.

Milošević, Slobodan. "The Ghosts of the Past, Present and Future: The Case of the Army Headquarters in Belgrade, Serbia." *Journal of Architecture and Urbanism* 39–1 (2015): 37–55.

Ministry of Defence of the Republic of Croatia, "Minister Krstičević Meets with the U.S. Secretary of Defense – Press Release." 13 July 2017. <www.morh.hr/en/news/press-releases/15027-minister-krsti%C4%8Devi%C4%87-meets-with-the-u-s-secretary-of-defense.html>.

Mitropoulos, Angela. "Archipelago of Risk: Uncertainty, Borders and Migration Detention Systems." *New Formations* 84–85 (2015): 163–183.

Mitropoulos, Angela and Matthew Kim. "Cross-Border Operations." *The New Inquiry*. 18 November 2015. <https://thenewinquiry.com/cross-border-operations/>.

Mitrovic, Olga. "Is the Use of Force Ever Legitimate in Order to Pursue Humanitarianism?" *Western Balkans Security Observer* 5:18 (2010): 79–90.

Močnik, Rastko. "Social Change in the Balkans." *Eurozine*. 20 March 2003. <www.eurozine.com/articles/2003-03-20-mocnik-en.html.gement>.

Močnik, Rastko. "The Balkans as an Elements in Ideological Mechanisms." In *Balkan as Metaphor*, edited by Dušan I. Bjelić and Obrad Savić, 79–119. Cambridge, MA: MIT Press, 2002.

Monbiot, George. "Consigning Their Future to Death." *The Guardian*. 22 April 1999.

Moorcock, Michael. "Dead God's Homecoming." In *Stormbringer*, 462–502. London: Millenium, 1993.

Morrison, David. "Kosovo: Lies, Half-Truths & Omissions." *Labour & Trade Union Review* (November 2000).

Moruzzi, Norma Claire. "National Abjects: Julia Kristeva on the Process of Political Self-Identification." In *Ethics, Politics, and Difference in Julia Kristeva's Writing*, edited by Oliver Kelly, 135–150. New York: Routledge, 1993.

Mostar Architects Association. "Mostar '92 – Urbicide." *Spazio e Societa'-Space and Society* 16:62 (1993): 8–25.

Mowrer, Paul Scott. *Balkanized Europe: A Study in Political Analysis and Reconstruction [1921]*. Ithaca: Cornell University Press, 2009.

Moynagh, Maureen. *Political Tourism and Its Texts*. Toronto: Toronto University Press, 2008.

Murphy, Sean D. *Humanitarian Intervention: The United Nations in an Evolving World Order*. Philadelphia: University of Pennsylvania Press, 1996.

Murphy, Sean D., Won Kidane and Thomas R. Snider. *Litigating War: Mass Civil Injury and the Eritrea-Ethiopia Claims Commission*. Oxford: Oxford University Press, 2013.

Naef, Patrick. "Tourism and the 'Martyred City': Memorializing War in the Former Yugoslavia." *Journal of Tourism and Cultural Change* 14:3 (2016): 222–239.

NATO. "NATO's Relations with Serbia." 11:59 (9 April 2015). <https://jfcnaples.nato.int/mlo_belgrade/about-mlo-belgrade/natos-relationship-with-serbia>.

NATO. *Helping Serbia Dispose of Stocks of Surplus Munitions*. 20 August 2013. <www.nato.int/cps/en/natolive/news_101894.htm?selectedLocale=en>.

NATO. *The Situation in and Around Kosovo* – Statement Issued at the Extraordinary Ministerial Meeting of the North Atlantic Council held at NATO Headquarters, Brussels, on 12 April 1999. <www.nato.int/docu/pr/1999/p99-051e.htm>.

NATO. *Press Statement by Dr. Javier Solana, Secretary General of NATO*, 23 March 1999. <www.nato.int/docu/pr/1999/p99-040e.htm>.

Nesković, Ivana. "Park Prijateljstva u Novom Beogradu." *Nasleđe* 12 (2011): 203–216.

Nietzsche, Friedrich. *The Gay Science*. New York: Dover Publications, 2006.

Nietzsche, Friedrich. *Twilight of the Idols and the Anti-Christ – Expeditions of an Untimely Man*. London: Penguin, 1990.

Nietzsche, Friedrich. *Beyond Good and Evil: Prelude to a Philosophy of the Future*. Harmondsworth, Middlesex: Penguin, 1973.

Nikolić, Zoran Lj. and Vidoje D. Golubović. *Beograd Ispod Beograda*. Beograd: Službeni List, 2006.

Nofi, Albert A. *Recent Trends in Thinking About Warfare*. Virginia: CNA Corporation, 2006.

Office of the United Nations High Commisioner for Human Rights. "Part III Methods and Means of Warfare Combatant and Prisoner-Of-War Status: Section I Method and Means of Warfare: Article 35(3)." In *Multilateral: Protocol additional to the Geneva Conventions of 12 August 1949, and relating to the protection of victims of international armed conflicts (Protocol I) (with annexes, Final Act of the Diplomatic Conference on the reaffirmation and development of international humanitarian law applicable in armed conflicts dated 10 June 1977 and resolutions adopted at the fourth session).*

Orwell, George. *Animal Farm*. London: Penguin, 2000.

Orwin, Clifford. "Humanitarian Military Intervention: Wars for the End of History?" *Social Philosophy & Policy Foundation* 23:1 (January 2006): 196–217.

Parthasarathy, D. "Decentralization, Pluralization, Balkanization? Challenges for Disaster Mitigation and Governance in Mumbai." *Habitat International* 52 (2016): 26–34.

Patton, Paul. "Conceptual Politics and the War Machine in Mille Plateaux." *SubStance* 13:3/4 (1984): 61–80.

Perović, Miloš R. *Iskustva Prošlosti*. Beograd: Zavod za planiranje razvoja grada Beograda, 1985.

Perović, Miloš R. and Spasoje Krunić, eds. *Nikola Dobrović: Eseji, Projekti, Kritike*. Beograd: Arhitektonski fakultet Univerziteta, Muzej arhitekture, 1998.

Polansky, Paul. *Unleaded Blood*. Kosovo: Kosovo Roma Refugee Foundation (KRRF), 2005.

Popović, Dragan. "Youth Labor Action (Omladinska Radna Akcija, ORA) as Ideological Holiday-Making." In *Yugoslavia's Sunny Side: A History of Tourism in Socialism (1950s–1980s)*, edited by Hannes Grandits and Karin Taylor, 279–302. Budapest: Central European Press, 2010.

Popović, Radmila. "Enemy Construction in the Media During the Kosovo Conflict: A Comparative View." *Graduate Student Workshop, Kokkalis Program on Southeastern and East Central Europe*. Harvard University: John F. Kennedy School of Government, 2001.

Portes, Alejandro and Robert L. Back. "Immigrant Earnings: Cuban and Mexican Immigrants in the United States." *International Migration Review* 14 (Fall 1980): 315–341.

Portes, Alejandro and Alex Stepick. "Unwelcome Immigrants: The Labor Market Experiences of 1980 (Mariel) Cuban and Haitian Refugees in South Florida." *American Sociological Review* 50 (August 1985): 493–514.

Pretes, Michael. "Tourism and Nationalism." *Annals of Tourism Research* 30:1 (2003): 124–142.

Prospect. "Roundtable: Francis Fukuyama, Eric Hobsbawm, Edward Luttwak, Timothy Garton Ash, Pierre Hassner and Robert Cooper Discuss Global Order in the 21st Century." 20 August 1999. <www.prospectmagazine.co.uk/magazine/theglobalorderinthe21stcentury>.

Ramadan, Adam. "Destroying Nahr el-Bared: Sovereignity and Urbicide in the Space of Exception." *Political Geography* 28:3 (2009): 153–163.

Ramzy, Austin. "U.N. Offers Regret but No Compensation for Kosovo Poisoning Victims." *New York Times*. 26 May 2017.

Rand, Ayn. *Global Balkanization*. Alviso: Palo Alto, 1977.

Raven, James. "Introduction: The Resonances of Loss." In *Lost Libraries: The Destruction of Great Book Collections Since Antiquity*, edited by James Raven, 1–41. New York: Palgrave Macmillan, 2004.

Ravetto, Kris. *The Unmaking of Fascist Aesthetics*. Minneapolis: University of Minnesota Press, 2001.

ReliefWeb. *UNEP/UNCHS (Habitat) Balkans Task Force (BTF) Situation Report No. 16.* 5 August 1999.

Riedlmayer, András J. "From the Ashes: "The Past and Future of Bosnia's Cultural Heritage." In *Islam and Bosnia: Conflict Resolution and Foreign Policy in Multi-Ethnic States*, edited by Maya Shatzmiller, 98–136. Montreal: McGill-Queens University Press, 2002.

Riegert, Kristina. *The Image War: NATO's Battle for Kosovo in the British Media.* Örebro: Örebro University, 2003.

Robinson, Guy and Alma Pobrić. "Nationalism and Identity in Post-Dayton Accords: Bosnia-Herzegovina." *Journal of Economic and Social Geography* 97:3 (2006): 237–252.

Roland, Ana and Bojana Jelovac. "Građani: Ne u NATO . . . ali se možda mora." *Blic.* 6 April 2015.

Ruchelman, Leonard I. *The World Trade Center – Politics and Policies of Skyscraper Development.* New York: Syracuse University Press, 1977.

Rynning, Sten. *NATO Renewed: The Power and Purpose of Transatlantic Cooperation.* New York: Palgrave Macmillan, 2005.

Safier, Michael. "Confronting 'Urbicide': Crimes Against Humanity, Civility and Diversity and the Case for a Civic-Cosmopolitan Response to the Attack on New York." *City* 5 (2001): 416–429.

Sandler, Tom and Keith Hartley. *The Political Economy of NATO: Past, Present, and into the 21st Century.* Cambridge: Cambridge University Press, 1999.

Sassen, Saskia. "Europe's Migrations: The Numbers and the Passions Are Not New." *Third Text* 20:6 (November 2006): 635–645.

Sassen, Saskia. *Guests and Aliens.* New York: The New Press, 2000.

Sassen, Saskia. *Losing Control?* New York: Columbia University Press, 1996.

Scarry, Elaine. *The Body in Pain: The Making and Unmaking of the World.* New York: Oxford University Press, 1985.

Schlamp, Hans-Jürgen (interviewer). "Clandestine Camps in Europe: Everyone Knew What Was Going in Bondsteel." *Der Spiegel.* Hamburg, 5 December 2005. <www.spiegel.de/international/clandestine-camps-in-europe-everyone-knew-what-was-going-on-in-bondsteel-a-388556.html>.

Schmitt, Carl. *Political Theology.* Chicago: University of Chicago Press, 2005.

Schwabach, Aaron. "Kosovo: Virtual War and International Law." *Law and Literature* 15:1 (2003): 1–21.

Shane, D. G. "Balkanization and the Postmodern City." In *Mortal City*, edited by Peter Lang. New York: Princeton Architectural Press, 58–64.

Shaw, Martin. "New Wars of the City: Relationships of 'Urbicide' and 'Genocide.'" In *Cities, War and Terrorism: Towards and Urban Geopolitics*, edited by Stephen Graham, 141–154. Oxford: Blackwell, 2004.

Simmel, Georg. "Bridge and Door." In *Rethinking Architecture*, edited by Neil Leach, 66–69. London: Routledge, 1997.

Simmons, Cynthia. "Urbicide and the Myth of Sarajevo." *Partisan Reviews* 68:4 (2001): 624–630.

Singer, P. W. *Wired for War.* New York: Penguin Press, 2009.

Sloterdijk, Peter. *Terror from the Air.* Cambridge, MA: MIT Press.

Smith, Neil. "New Globalism, New Urbanism: Gentrification as Global Urban Strategy." *Antipode* 34:3 (July 2002): 427–450.

Sorensen, Jens Stilhoff. *State Collapse and Reconstruction in the Periphery: Political Economy, Ethnicity and Development in Yugoslavia, Serbia and Kosovo.* New York: Berghahn Books, 1999.

Sorkin, Michael. "Urban Warfare: A Tour of the Battlefield." In *Cities, War and Terrorism: Towards and Urban Geopolitics*, edited by Stephen Graham, 251–263. Oxford: Blackwell, 2004.

Stanely, Bruce. "The City-Logic of Resistance: Subverting Urbicide in the Middle East City." *Journal of Peacekeeping and Development* 27:3 (2013): 10–24.

Štitkovac, Ejub. "Croatia: The First War." In *Burn This House: The Making and Unmaking of Yugoslavia*, edited by Jasminka Udovicki and James Ridgeway, 154–175. Durham: Duke University Press, 1997.

Stoker, Bram. *Dracula*. Westminster: Archibald Constable and Company, 1897.

Suhrke, Astri, Michael Barutciski, Peta Sandison and Rick Garlock. "The Kosovo Refugee Crisis." *United Nations High Commissioner for Refugees Evaluation and Policy Analysis Unit*. Switzerland: Geneva, 2000. <https://www.unhcr.org/3ba0bbeb4.pdf>.

Swiffen, Amy. *Law, Ethics and the Biopolitical*. New York: Routledge, 2010.

Taylor, Philip M. *Global Communications, International Affairs and the Media Since 1945*. London: Routledge, 1997.

Tester, Keith. "The Terminal Paradox of Globalization." In *Globalization, Political Violence and Translation*, edited by Esperanza Bielsa and Christopher W. Hughes, 25–44. New York: Palgrave Macmillan, 2009.

Todorova, Maria. *Imagining the Balkans*. New York: Oxford University Press, 2009.

Todorova, Maria. "The Trap of Backwardness: Modernity, Temporality, and the Study of Eastern European Nationalism." *Slavic Review* 64:1 (2005): 140–164.

Todorova, Maria. "The Balkans: From Discovery to Invention." *Slavic Review* 53:2 (1994): 453–482.

Tolbert, Julian H. "Crony Attack. Strategic Attack's Silver Bullet?" Masters Dissertation, School of Advanced Air and Space Studies, Alabama 2006.

Tourneur, Jacques, dir. *Cat People. Film*. US: RKO Radio Pictures Inc, 1942.

Tschumi, Bernar. *Architecture Concepts: Red Is Not a Color*. New York: Rizzoli International Publications, 2012.

UNEP/UNCHS. *The Kosovo Conflict – Consequences for the Environment & Human Settlements*. <http://wedocs.unep.org/bitstream/handle/20.500.11822/8433/-The%20Kosovo%20Conflict%20Consequences%20for%20the%20Environment%20%26%20Human%20Settlements-1999378.pdf?sequence=3&isAllowed=y>.

UNESCO. "Report on the Status of Cultural Heritage in the Old Town of Dubrovnik following the bombardments of October, November and December 1991." In *Report of the UNESCO Mission to Dubrovnik 27 November – 22 December 1991*. <www.heritage.sense-agency.com/assets/dubrovnik-old-city/sg-2-05-unesco-report-en.pdf>.

UNHCR, "Figures at a Glance," 2018, < www.unhcr.org/figures-at-a-glance.html>.

United States Institute of Peace. *Interim Agreement for Peace and Self-Government in Kosovo Rambouillet*. France, 23 February 1999. < https://www.usip.org/sites/default/files/file/resources/collections/peace_agreements/kosovo_ramb.pdf>.

Vasić, Biljana. "Postavljanje spomen-obeliska u Beogradu." *Vreme 493*. 17 June 2000.

Vaux, Tony. *The Selfish Altruist*. London: Earthscan Publishing, 2001.

Venturi, Robert et al. *Learning from Las Vegas: The Forgotten Symbolism of Architectural Form*. Cambridge, MA: MIT Press, 1977.

Vilenica, Ana. "The Art of New Class Geography of the City: Culture-Guided Urban Regeneration Serving the Modernization of the Periphery." In *The Grey Zones of Creativity and Capital*, edited by Gordana Nikolić and Šefik Tatlić. Amsterdam: Institute of Networked Cultures, 2015.

Virilio, Paul. *Speed and Politics: An Essay on Dromology*. Los Angeles: Semiotext(e), 2006.

Virilio, Paul. *A Landscape of Events*. Cambridge, MA: MIT Press, 2000.

Virilio, Paul. *Bunker Archaeology*. New York: Princeton Architectural Press, 1994.

Virilio, Paul. *War and Cinema: The Logistics of Perception*. London: Verso, 1989.

Virilio, Paul. *Popular Defense and Ecological Struggles*. New York: Semiotext(e), 1990.

Virilio, Paul and John Armitage. "Ctheory Interview with Paul Virilio: The Kosovo War Took Place in Orbital Space." *ctheory*. 2000. <https://journals.uvic.ca/index.php/ctheory/article/view/14599>.

Virilio, Paul and Sylvère Lotringer. *Crepsular Dawn*. New York: Semiotext(e), 2002.

Vöckler, Kai. " 'Balkanology' and the Future of the European City." In *Balkanology – SAM No. 6*, edited by SAM, Francesca Ferguson and Kai Vockler. Basel: Christoph Merian Verlag, 2008.

Von Eckardt, Wolf. "New York's Trade Center: World's Tallest Fiasco." *Harper's Magazine* (1966): 94–100.

Vukmirović, Zorka B. "Release of Hazardous, Toxic and Cancerogenic Substances into the Atmosphere During the Warfare in Yugoslavia and an Assessment of Their Environmental Impact." *Environmental Studies, Alternative Academic Educational Network. Belgrade*, 26 October 1999.

Vuksanović-Macura, Zlata. *Život na Ivici*. Beograd: Orion Art, 2012.

VU University Amsterdam. *Human Costs of Border Control*. <www.borderdeaths.org>.

Wachtel, Andrew Baruch. *The Balkans in World History*. New York: Oxford University Press, 2008.

Wacquant, Loïc. "The Militarization of Urban Marginality: Lessons from the Brazilian Metropolis." *International Political Sociology* 2:1 (2008): 56–74.

Wacquiant, Loic. *Urban Outcasts*. Cambridge: Polity, 2007.

Watson, Iain. "(Re) Constructing a World City: Urbicide in Global Korea." *Globalizations* 10:2 (2013): 309–325.

Wear, Andrew. *Knowledge and Practice in English Medicine, 1550–1680*. Cambridge: Cambridge University Press, 2000.

Weber, Samuel. *Targets of Opportunity: On the Militarization of Thinking*. New York: Fordham University Press, 2005.

Weber, Samuel. "War, Terrorism, and Spectacle, or: On Towers and Caves." *Grey Room* 7 (2002): 14–23.

Weiss, Srdjan Jovanović. "National, Un-National." *Nationalities Papers: The Journal of Nationalism and Ethnicity* 41:1 (2013): 90–108.

Weiss, Srdjan Jovanović. "Milosevic as Architect." In *Networked Cultures: Parallel Architectures and the Politics of Space*, edited by Helge Mooshammer and Peter Mörtenböck. Rotterdam: NAi Publishers, 2008.

Weiss, Srdjan Jovanović. *Almost Architecture*. Stuttgart: Merz+Solitude, 2006.

Weiss, Srdjan Jovanović. "School of Missing Studies." In *Did Someone Say Participate? An Atlas of Spatial Practice*, edited by Markus Miessen and Shumon Basar, 241–257. Cambridge, MA: MIT Press, 2006.

Weiss, Srdjan Jovanović. "NATO as Architectural Critic." *Cabinet* 1 (2000/01). <www.cabinetmagazine.org/issues/1/NATO.php>.

Weizman, Eyal. *Forensic Architecture: Notes from Fields and Forums/Forensische Architectur. Notizen von Felden und Foren, dOCUMENTA (13)*. Ostfildern, Germany: Hatje Cantz, 2012.

Weizman, Eyal. *The Least of All Possible Evils: Humanitarian Violence from Arendt to Gaza*. London: Verso, 2011.

Weizman, Eyal. *Hollow Land: Israel's Architecture of Occupation*. London: Verso, 2007.

Wenders, Wim and Mary Zournazi. *Inventing Peace: A Dialogue of Perception*. London: I.B.Tauris, 2013.

West, Rebecca. *Black Lamb and Grey Falcon: A Journey Through Yugoslavia*. Edinburgh: Canongate Books, 2006.

Wheeler, Nicholas J. *Saving Strangers: Humanitarian Intervention in International Society*. Oxford: Oxford University Press, 2000.

Wieseltier, Leon. "Force Without Force." *New Republic*. 26 April, 1999.

Wilmer, Franke. *The Social Construction of Man, the State, and War: Identity, Conflict and Violence in Former Yugoslavia*. New York: Routledge, 2002.

Wolfson, Adam. "How to Think About Humanitarian War." *Commentary* 108:7 (1 July 2000). <www.commentarymagazine.com/articles/how-to-think-about-humanitarian-war/>.

Woodward Jr., John D. "The Law and the Use of Biometrics." In *Handbook of Biometrics*, edited by Anil K. Jain et al., 357–379. New York: Springer, 2007.

Woodward, Susan. *Balkan Tragedy: Chaos and Dissolution After the Cold War*. Washington, DC: Brookings Institution, 1995.

World Health Organization Report. *Depleted Uranium Sources, Exposure and Health Effects*, April 2001. <http://apps.who.int/iris/bitstream/handle/10665/66930/WHO_SDE_PHE_01.1.pdf?sequence=1&isAllowed=y>.

World Health Organization Report. *Report of the World Health Organization Depleted Uranium Mission to Kosovo – Undertaken at the request of the Special Representative of the Secretary-General and Head of the United Nations Interim Administration Mission in Kosovo* (UNMIK), *22–31 January 2001*. <http://apps.who.int/iris/handle/10665/108666>.

Xhunga, Marjola. "Integration in Albania: No longer a pipe dream," UNHCR, 14 April 2003. < https://www.unhcr.org/news/latest/2003/4/3e9ab9a84/integration-albania-longer-pipe-dream.html>.

Yarrow, Ruth. "The Continuing War in Yugoslavia: Environmental Effects." *Peacework*. September 1999.

Yeomans, Rory. "How to Combat Croatian Revisionists' Culture of Lies." *BalkanInsight*. 6 June 2018. <www.balkaninsight.com/en/article/how-to-combat-croatian-revisionists-culture-of-lies-06-05-2018>.

Yeomans, Rory. *The Utopia of Terror*. Rochester: University of Rochester Press, 2015.

Yeomans, Rory. *Visions of Annihilation*. Pittsburgh: University of Pittsburgh Press, 2013.

Žaknić, Ivan. *Journey to the East*. Cambridge, MA: MIT Press, 2007.

Zolberg, Aristide R. and Long Litt Woon. "Why Islam Is Like Spanish: Cultural Incorporation in Europe and the United States." *Politics and Society* 27:1 (March 1999): 5–38.

Zwijnenburg, Wim. "Hazard Aware: Lessons Learned from Military Field Manuals on Depleted Uranium and How to Move Forward for Civilian Protection Norms." Report. Netherlands: PAX, 2012.

Index

Note: Page numbers in *italic* indicate a figure on the corresponding page.